DNA Technology

Books in the **Contemporary World Issues** series address vital issues in today's society such as genetic engineering, pollution, and biodiversity. Written by professional writers, scholars, and nonacademic experts, these books are authoritative, clearly written, up-to-date, and objective. They provide a good starting point for research by high school and college students, scholars, and general readers as well as by legislators, businesspeople, activists, and others.

Each book, carefully organized and easy to use, contains an overview of the subject, a detailed chronology, biographical sketches, facts and data and/or documents and other primary source material, a forum of authoritative perspective essays, annotated lists of print and nonprint resources, and an index.

Readers of books in the Contemporary World Issues series will find the information they need in order to have a better understanding of the social, political, environmental, and economic issues facing the world today.

CONTEMPORARY WORLD ISSUES

DNA Technology

A REFERENCE HANDBOOK

Second Edition

David E. Newton

 ABC-CLIO™

An Imprint of ABC-CLIO, LLC
Santa Barbara, California • Denver, Colorado

Library of Congress Cataloging-in-Publication Data

Names: Newton, David E., author.
Title: DNA technology : a reference handbook /
    David E. Newton.
Other titles: Contemporary world issues.
Description: Second edition. | Santa Barbara, California :
    ABC-CLIO, [2017] | Series: Contemporary world
    issues | Includes bibliographical references and index.
Identifiers: LCCN 2016029735 (print) | LCCN 2016030934
    (ebook) | ISBN 9781440850479 (alk. paper) |
    ISBN 9781440850486 (ebook)
Subjects: | MESH: Genetic Techniques | DNA |
    Biotechnology | Genetics
Classification: LCC QH442 (print) | LCC QH442 (ebook) |
    NLM QU 550 | DDC 660.6/5—dc23
LC record available at https://lccn.loc.gov/2016029735

ISBN: 978-1-4408-5047-9
EISBN: 978-1-4408-5048-6

21 20 19 18 17   1 2 3 4 5

This book is also available as an eBook.

ABC-CLIO
An Imprint of ABC-CLIO, LLC

ABC-CLIO, LLC
130 Cremona Drive, P.O. Box 1911
Santa Barbara, California 93116-1911
www.abc-clio.com

This book is printed on acid-free paper ∞

Manufactured in the United States of America

Contents

On April 25, 1953, American biologist James Watson and English chemist Francis Crick published a paper in the British journal *Nature* proposing a chemical structure for a family of chemical compounds known as deoxyribonucleic acid (DNA). The structure they suggested consisted of a long pair of chains twisted around each other, somewhat like a spiral staircase, with subunits called nitrogen bases paired with each other as steps on the staircase. At the conclusion of their paper, Watson and Crick noted, "It has not escaped our notice that the specific pairing we have postulated immediately suggests a possible copying mechanism for the genetic material." What Watson and Crick had discovered is that DNA molecules carry a "code" for genetic characteristics, such as hair and skin color, ear shape, color of eyes, height, and a host of other traits. This discovery solved a problem that had troubled scientists for decades: How does like produce like during the process of reproduction? How does a willow tree always produce other willow trees only, and never oak trees? How do redheaded children appear only in families with redheaded parents, grandparents, and/or great-grandparents, and never in families whose predecessors had only black or brown hair?

The very remarkable thing about the Watson and Crick discovery is that the answer to these and other biological questions was given in chemical terms. After Watson and Crick's paper, scientists were able to think about how hereditary traits are transmitted in terms of atoms and molecules. And, since chemists already knew a great deal about manipulating atoms and molecules, the possibility presented itself that biologists might now be able to manipulate the characteristics of living organisms in much the same way: by moving around atoms and molecules.

Today, because of this discovery, biological studies are as much a function of chemical changes as are industrial processes, such as the production of paper, the manufacture of plastics, and the processing of metallic materials.

An even more profound consequence of the Watson and Crick discovery is that one can now say that, to a very considerable extent, it is possible to describe everything we know about the biological characteristics of a human being (or a chimpanzee or elephant or flea) by writing chemical formulas and chemical equations. To the average person, the following expression may just be a series of letters, or, to someone with an introductory course in chemistry, a series of nitrogen bases in a DNA molecule. But to a molecular biologist, that string of letters stands for some physical trait, such as hair color or a potential health problem, such as sickle-cell anemia.

This possibility was apparent to scientists from almost the moment that Watson and Crick published their 1953 paper. Researchers began to ask if it might be possible to alter the "genetic code" in an amoeba, or a fruit fly, or even a human being in order to change that organism's genetic composition and, therefore, its physical characteristics. Out of that line of thinking has come a whole new field of study that uses information about DNA structure to develop practical applications. This book reviews those applications in six fields:

forensic science, the development of transgenic organisms (organisms with genetic characteristics of two different species), genetic testing, gene therapy, molecular farming (also known as pharming), and cloning.

Perhaps the most remarkable feature of the history of DNA technology is the tempo at which progress has been made. Only 50 years after the Watson and Crick discovery, scientists are applying that knowledge to identifying genetic diseases in fetuses, altering tobacco plants so that they will produce useful drugs, identifying criminals accused of crimes with a certainty of one in a billion or better, treating (and, perhaps, curing) adults with life-threatening genetic disorders, and producing multiple copies (clones) of animals born outside the normal process of reproduction.

Chapter 1 provides a historical review of the development of these six DNA technologies, with an overview of the scientific principles fundamental to each technology. Chapter 2 reviews some social, ethical, political, and other issues that have arisen as a result of the development of these technologies. In many cases, the use of DNA technology has raised some fundamental philosophical and theological questions about what it means to be human, or what it means to be alive. One should not be surprised that these questions engender serious, and often heated, debates as to the conditions under which a particular technology should be used or, indeed, if it should ever be used at all. Chapter 3 focuses on the international aspects of the debates over DNA technology. Perhaps the most significant findings of this chapter are the enormous diversity in the use of DNA technologies in various parts of the world, the extent to which these technologies have stirred up controversy over their use, and the range of views that people from different parts of the world have about DNA technology.

Chapters 4 through 8 provide a variety of resources to help the reader continue a study of DNA technology. Chapter 4 offers a chronology of important events in the growth of scientific knowledge about the nature of DNA and its role in biological processes, along with some important political, social, legal, and other events related to this growth of knowledge and its applications in everyday life. Chapter 5 provides brief biological sketches of some important figures in the development of DNA technology and its applications. Chapter 6 contains some political and legal documents that illustrate the way the United States and other countries around the world are attempting to assess, regulate, and control the applications of DNA technology in a variety of fields. Some relevant data about the growth of this field of study are also included. Chapter 7 provides a list of organizations interested in DNA technology, some whose primary responsibility is regulatory, some interested in advancing the development and use of DNA technology, and some working to moderate or stop the development of that field of research. Chapter 8 is an annotated bibliography of print and electronic resources for the six major types of DNA technology discussed in the book. The book concludes with a glossary of important terms used in the field of DNA technology.

Preface to the Second Edition

It is generally difficult to predict the precise direction of any particular subject in the sciences. Unexpected discoveries—such as the determination of the three-dimensional structure of the DNA molecule by Watson and Crick in 1953—may be both profound surprising and, at the same time, gateways to entirely new directions in the future of a scientific field. Such a statement is as true for the technology of DNA research as it is for any other field of science. The discovery of a mechanism for gene editing in a relatively simple and straightforward manner by Emmanuelle Charpentier and Jennifer Doudna in 2012 was both a startling breakthrough in the field of DNA technology and, at the same time, a tool for further developments in the field that researchers had only been able to hope for only a few years earlier.

For this reason, any book on the subject of DNA technology faces the challenge of continuously updating to stay up with changes in the field, new applications that have become available, and evolving or new social, economic, technological, political, economic, and other issues that may arise because of those breakthroughs. The second edition of *DNA Technology: A Reference Handbook* is our effort, then, to meet that challenge. The book attempts, in the first place, to lay out for readers some of the most important scientific and technical changes that have occurred since publication of the first edition, such as the discovery and use of gene-editing techniques such as

CRISPR-Cas9, zinc finger nucleases, and other types of gene-editing protocols, and the developing significance of low copy number DNA technology, along with the everyday challenges create by such technologies.

At the same time, this update is designed to keep readers up-to-date on developments in somewhat older technologies and issues, such as the development of genetically modified crops, animals, and foods, and the applications of DNA analysis in the forensic sciences. Topics such as these were discussed in some detail in the first edition of the book, but the debate as to how they are to be used and how society will be affected by them continues in actions such as the 2014 Vermont law on the labeling of genetically modified foods.

The backbone of the discussion of DNA technology presented in the first edition of the book, then, remains with this edition. But important changes that have occurred in the field are described and discussed both in the introductory chapters of the book and in the resources presented in Chapter 4 (Profiles), Chapter 5 (Data and Documents), Annotated Bibliography (Chapter 6), and Chronology (Chapter 7). In addition, a new Chapter 3, titled "Perspectives," provides interested individuals with an opportunity to discuss in greater detail some very specific issues within the general area of DNA technology.

DNA Technology

Introduction

Are you in the market for a new pet? How about a fish that glows in the dark? Even under white light, this new GloFish® is a striking shade of red, green, or yellow; you get to choose from Starfire Red®, Electric Green®, or Sunburst Orange®. GloFish® are one of the somewhat unexpected products of DNA technology. They were first invented in 1999 by scientists at the National University of Singapore who were trying to develop a fish that could be used to warn of the presence of pollutants in water. They added a gene that codes for a protein known as green fluorescent protein (GFP) that occurs naturally in jellyfish. Their plan was to further program the fish to display its color when a particular pollutant was present in the water. While the research was still under way, entrepreneurs from a U.S. company, Yorktown Technologies, L.P., met with inventors of the GloFish® to work out rights for licensing and selling the engineered fish in the United States. After receiving approval from all relevant U.S. regulatory agencies, Yorktown Technologies began selling GloFish® in late 2003. The age of engineered life forms has arrived in your living room!

GloFish swim in an aquarium at a pet store in Alexandria, Virginia. The trademarked fish is genetically engineered. (Mark Wilson/Getty Images)

Modifying Life: The Early History

Humans around the world today have a host of plants and animals at their disposal as a source of food, to help in daily work, to serve as helping animals and pets, and for countless other purposes. They range from strawberries and corn and cabbage to horses and dogs and sheep. All of these organisms have evolved over long periods of time from one or more primitive forms. Sometimes evolution has occurred naturally, without any efforts on the part of humans. For example, two closely related species may mate by one method or another and a new, slightly different species is produced. On other occasions, humans have attempted to breed organisms with each other in order to produce plants and animals that are healthier, better tasting, more attractive, stronger, or better in some other way than their progenitors. These efforts resulted in the domestication of many plants and animals long before the period of written history: by 13000 BCE, in the case of rice; by 10000 BCE in the case of dogs, goats, pigs, and sheep; by 7000 BCE in the case of the potato; and by 2700 BCE in the case of beans, chilies, corn, and squash (Macrohistory: World History 2016). Evidence for the hand pollination of plants—date palms—dates to the Assyrians and Babylonians in about 700 BCE (History of Plant Breeding 2015). In many cases, humans controlled the reproduction of plants and animals without any knowledge of the scientific principles involved; they simply applied information gained from centuries of experience based on trial-and-error experiments. They bred sheep with the best wool, potatoes with the best taste, horses with the greatest strength, corn that was most nutritious, and other plants and animals with desirable properties. Without understanding how, they directed the evolutionary paths of untold numbers of plant and animal species.

In some cases, they even produced completely new plants and animals by hybridization (the crossbreeding of different species or varieties of plants or animals). One of the earliest examples may have been the mule and the hinny. A mule is the offspring of a male donkey (a jackass) and a female horse (a mare), while a hinny is the offspring of a female donkey (jenny) and a male horse (a stallion). Archaeologists have found representations of mules and hinnies in Egyptian tomb paintings dating back to at least 1400 BCE (Sherman 2002, 42). Throughout the centuries, humans have produced a number of other hybrid animals in much the same way the mule and hinny must first have appeared: the beefalo (a cross between an American bison and a domestic cow), Bengal cat (domestic cat and Asian leopard cat), zeedonk (donkey and zebra), dzo (domestic cow and yak), Leyland cypress (Monterey cypress and Nootka cypress), peppermint (water mint and spearmint), and loganberry (raspberry and blackberry). Although most of the details of the ways in which humans have influenced the evolution of plants and animals by hybridization are unavailable (because they occurred so long ago), it is clear that that influence has been profound: plants and animals in the 21st century are very much the product of human modification.

The Birth of Genetics

By the last quarter of the 17th century, hybridization had taken on a new, more scientific flavor, with breeders and farmers using a more rational approach to the crossing of species. Horticulturists in the Netherlands were a particularly good example, having found dependable methods for producing a large variety of new plants, including the first hybrid hyacinth and a number of new tulip varieties (Murphy 2007, 246, 331). Without much doubt, however, the first true research designed to produce a scientific understanding of the process by which genetic traits are passed from one generation to the next was conducted by the Austrian monk Gregor Mendel between 1856 and 1863. Mendel brought to his research a valuable combination of interests in the evolution of organisms, natural history, and the physical sciences. He studied seven physical properties of the common pea plant (*Pisum sativum*): flower color (purple or white), flower position (axial or terminal), stem length (short or long), seed shape (round or wrinkled), seed color (yellow or green), pod shape (constricted or inflated), and pod color (yellow or green). His experiments were relatively straightforward. Pea plants with various combinations of these traits were crossbred with each other, and the characteristics of offspring were noted. The offspring were then crossbred with each other, and characteristics of the next generation were also recorded. In this way, the transmission of traits from parents to the first generation to the second generation, and so on, could be observed. Mendel found, as one simple example, that the cross between a pea plant with yellow seeds and a plant with green seeds produces offspring, all of whom have yellow seeds only. If the offspring (the f1 generation) are then crossbred among themselves, plants with green seeds reappear in one out of four cases in the next (f2) generation. Further, the 3:1 ratio of traits continues to reappear in future (f3, f4, etc.) generations (for an excellent overview of Mendel's work, see Mawer 2006).

Mendel's work is significant for a number of reasons. First, it contradicted the belief at his time that genetic traits tend to blend in the passage of generations. That is, classic theory at the time would have predicted that the cross described previously would have resulted in pea plants with greenish-yellow seeds in some later generation, with the shade of green-yellow varying from plant to plant. Second, it suggested that the transmission of hereditary characteristics requires the presence of some kind of "carrying unit" for a genetic trait. Mendel referred to these units as *factors, unit factors*, or, simply, *units*, but he really had no idea as to what these units were. Third, Mendel demonstrated that an organism might possess the unit factor for a trait, but might not actually show that trait. In the previous example, for example, pea plants in the f1 generation all had yellow seeds. But they must somehow have retained the unit factor for green color, because in the next generation one out of four plants had green seeds. Based on these observations, Mendel also concluded that each offspring plant must receive one unit factor from each parent. Finally, he hypothesized that, for each trait, one unit factor must be dominant, and the other recessive. That is, plants that inherited both yellow and green unit factors actually produced only yellow seeds, so the yellow factor must be dominant over the green factor. Only plants that received two green unit factors from parents could produce green seeds.

Mendel reported the results of his research at two meetings of the Natural History Society of Brno (also Brünn) on February 8 and March 8, 1865. He then published the text of his remarks in a paper, "Versuche über Pflanzen-Hybriden" ("Experiments on Plant Hybridization"), in the Proceedings of the Natural History Society of Brunn for the year 1865 (published in 1866). His verbal and written reports were almost entirely ignored. Over the next 35 years, his work was cited only a handful of times, and it had essentially no effect on biological thought. Only in 1900 did Mendelian genetics once more see the light of day when three biologists (Hugo de Vries, Carl Correns, and Erich von Tschermak), all working independently of each other, rediscovered Mendel's work and brought that work to the attention of the biological community. Mendel had simply been born three decades too early!

For more than a century, Mendelian genetics has been a powerful tool for understanding the transmission of hereditary traits. Geneticists now deal with much more complex systems than those used by Mendel, but the principles he elucidated, modified and updated continue to explain most of the basic facts of inherited traits. The one important problem that remained unsolved for decades after Mendel's rediscovery in 1900 was the nature of the transmitting unit, the unit factor that Mendel invented to explain his findings. Later biologists sometimes invented new names for the unit factor. Hugo de Vries himself suggested the term *pangen* in 1889, long before he knew of Mendel's work, and the modern name for the unit factor—*gene*—was invented in 1903 by Danish botanist and geneticist Wilhelm Johannsen. The fundamental problem was that no one really had any idea as to what a unit factor (or pangen or gene) was. The naming game was simply a method for identifying a "black box" in organisms responsible for transmitting genetic traits.

In 1953, that problem was finally resolved. American biologist James D. Watson and British chemist Francis Crick discovered that the "black box" is a chemical compound, deoxyribonucleic acid, or DNA.

The Road to DNA

Like virtually all scientific breakthroughs, the discovery of DNA did not spring fully formed from Watson's and Crick's brains. It was the result of decades of research and theorizing about the nature of the genetic material. DNA had been discovered as far back as 1869 by Swiss physician and biologist Johannes Friedrich Miescher, although Miescher knew of no biological role for the compound (which he called *nuclein*). In fact, scientists knew very little about the structure or function of DNA for nearly seven decades after its discovery. For most of that time, biologists were convinced that unit factors (or genes) must consist of some kind of protein-like material. Proteins are very large, complex molecules consisting of long chains of amino acids. By "very large," one means molecules contain tens of thousands, hundreds of thousands, or millions of atoms. It made sense to believe that molecules of such complexity would be the ideal carriers of complex genetic information. By contrast, DNA is a relatively simple molecule consisting of only three basic units: a sugar (deoxyribose), a phosphate group, and one of four nitrogen bases.

One of the first clues in the DNA story came in 1909, when Russian American chemist Phoebus Levene discovered that one type of nucleic acid contains a sugar called *ribose*. Levene called that type of nucleic acid *ribonucleic acid*, or *RNA*. Twenty years later, Levene found a second sugar in other types of nucleic acid, the sugar deoxyribose. Deoxyribose differs from ribose only in that it contains one less hydroxyl group (-OH), thus the name, *de-* ("without") *-oxy-* ("oxygen") ribose. To these nucleic acids he gave the name *deoxyribonucleic acid*, or *DNA*. A hint that some form of nucleic acid might be involved in heredity came in 1924 when researchers found that both protein and nucleic acids were present in chromosomes. This information was essential for the identification of genes, of course, because biologists already knew that genes are located on chromosomes. This discovery meant that both proteins and nucleic acids were candidates for the role of genes. It did not tell which family of compounds made up genes, and scientists continued to favor a protein hypothesis.

Another clue to the puzzle came in 1928 when British medical officer Franklin Griffith found that the hereditary material (genes) was something that was stable when heated, a characteristic not often common with proteins. Griffith killed bacteria by heating them and then inserted their cellular components into live bacteria. He found that genetic information from the killed bacteria was expressed within the living bacteria. The answer as to which heat-resistant compound it was that produced this result was answered in 1944 in a series of elegant and relatively simple experiments conducted by Oswald Avery, Colin MacLeod, and Maclyn McCarty. The three researchers worked with two types of pneumococci bacteria (the bacteria that cause pneumonia), identified as the R-type (for "rough" coating) and S-type (for "smooth" coating). They removed from the nuclei of S-type bacteria all protein, DNA, and other compounds and transferred each type of compound to R-type bacteria. Of the host bacteria, only those that received DNA were transformed into S-type bacteria. The results were strongly suggestive that DNA was the hereditary material; that is, that genes consisted of DNA in some form. Strangely enough, most researchers were reluctant to accept the implications of the Avery-MacLeod-McCarty experiment (which may explain in part why none of the three ever received a Nobel Prize).

One more piece in the DNA puzzle was put in place in 1949 when Austrian American Erwin Chargaff discovered that the amount of adenine in a DNA molecule is always the same as the amount of thymine, and the amount of cytosine, the same as the amount of guanine. Adenine, cytosine, guanine, and thymine are four nitrogen bases that make up DNA molecules. Chargaff apparently saw no significance for DNA in his results, but Watson and Crick later did.

By the early 1950s, most of the clues needed for the deciphering of the DNA structure were in place. The key figures in the research were almost within shouting distance of each other: Watson and Crick at the Cavendish Laboratory at the University of Cambridge, and Maurice Wilkins and Rosalind Franklin at King's College, London. While Watson and Crick were manipulating the information they had about DNA from Avery, Chagraff, and other researchers, Wilkins and Franklin were taking X-ray photographs of the DNA molecules. X-ray crystallography is a very powerful, but very difficult, technique for determining the structure of complex molecules by shining X-rays through them. The images obtained by this method are often very difficult to interpret and to translate into three-dimensional models of the molecules. By late 1952, Franklin had obtained extraordinary images of DNA molecules, which appeared to provide the final clue to the compound's structure. She hesitated in announcing her results, however, wanting to further confirm them. Crick and Watson, who had seen her results, were not so shy. They realized that Franklin's images gave them the last bit of information they needed, and by March 7, 1953, had constructed their model of the DNA molecule. They reported the results of their research in a paper that appeared in

the April 25, 1953, issue of the journal *Nature*, titled "A Structure for Deoxyribose Nucleic Acid." They began their paper
with the comment:

> We wish to suggest a structure for the salt of deoxyribose
> nucleic acid (D.N.A.). This structure has novel features
> which are of considerable biological interest. (Watson and
> Crick 1953, 737–738)

Watson and Crick (with help from Franklin and others) had
unraveled the structure of the DNA molecule. The gene had
finally been given a clear and unequivocal chemical and physical structure.

The Structure of DNA

So what did Watson and Crick discover? They found that a
DNA molecule consists of two very long spaghetti-like strands
(the "backbone" of the molecule) made of alternating sugar
(deoxyribose) and phosphate groups (see Figure 1.1).

These two strands are wrapped around each other, a bit like a
circular staircase, in a geometric form known as a *double helix*.

S - P - S - P - S - P - S - P - S - P - S - P - S - P - S - P - S - P - S - P - S - P - S -.

Figure 1.1   Backbone of DNA Molecule

S - P - S - P - S - P - S - P - S - P - S - P - S - P - S - P - S - P - S - P - S - P - S -
I   I   I   I   I   I   I   I   I   I   I
C   A   C   T   G   G   G   C   A   T   G   C

Figure 1.2   Nitrogen Bases in DNA
(The placement of nitrogen bases is random in the molecule.)

Attached to each sugar molecule is one of four nitrogen bases: adenine, cytosine, guanine, or thymine, as shown in Figure 1.2. The combination of one sugar (deoxyribose) unit and one nitrogen base, as in S–A, is called a *nucleoside*; the combination of a sugar, base, and phosphate group, as shown in Figure 1.3 is known as a *nucleotide*.

S - A
|
P

Figure 1.3   Nucleotide

To a chemist, then, the long string that makes up a DNA backbone is a *polynucleotide, many* ("poly-") nucleotide units joined to each other.

The two strands are then wrapped around each other in such a way that the nitrogen bases are inside the molecule, with complementary bases adjacent to each other, as shown in Figure 1.4.

(The placement of nitrogen bases is random in the molecule.)

In this structure, the nitrogen bases are paired with each other so that every adenine is joined to a thymine, and every cytosine to a guanine. How this structure translates into genetic information is discussed later in this chapter.

At this point, a word needs to be said about DNA's perhaps somewhat less well-known cousin, RNA. The two families of compounds are similar in many ways, especially with regard to their importance in the transfer of genetic information from one generation to the next. But structurally, they differ in three important ways. First of all, the sugar in the backbone of an RNA molecule is ribose, not deoxyribose. Second, the four nitrogen bases in RNA are adenine, cytosine, guanine, and uracil, rather than thymine. Third, RNA is a single-stranded molecule and not a double-stranded helix, as in the case of DNA. A typical RNA molecule, then, might have a structure like that shown in Figure 1.5.

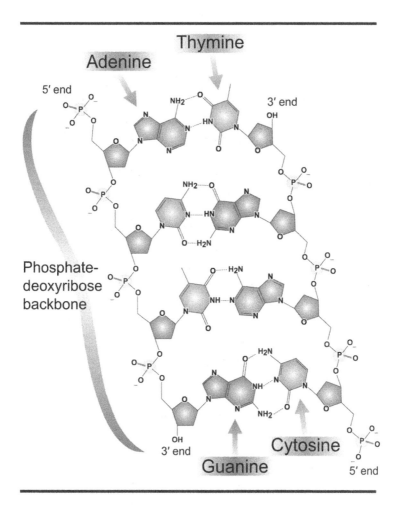

Figure 1.4   Double-Helix Structure of the DNA Molecule

```
- P - R - P - R - P - R - P - R - P - R - P - R - P - R - P - R - P - R - P - R - P - R -
      |       |       |       |       |       |       |       |       |       |       |
      A       C       U       G       G       U       C       C       C       G       U       U
```

Figure 1.5   Structure of an RNA Molecule

The Rise of Molecular Genetics

The discovery of DNA structure made by Watson and Crick is sometimes cited as being one of the most important discoveries in the history of science. The reason for this accolade is that prior to the discovery, geneticists knew a great deal about the way genetic traits were transmitted from generation to generation. But they knew nothing at all as to how humans might manipulate that process at the most fundamental level, that of cells and molecules. Beyond trial-and-error hybridization experiments, genetic manipulation was essentially a mystery. After Watson and Crick, scientists realized that the genetic material consisted of molecules—molecules of DNA—that could (at least in theory) be synthesized, broken apart, and manipulated in essentially the same way as any other kind of chemical molecule. The opportunities for engineering genes and altering heredity seemed endless.

Before the "at least in theory" could become a practical reality, however, a great many technical problems for working with DNA molecules needed to be solved. Francis Crick was intimately involved in solving the first two of those problems, the mechanism by which genetic information is coded in a DNA molecule, and the process by which that information is used to synthesize proteins in a cell. In a paper published in 1958, Crick stated two general principles about the function of DNA. The first, which he called the Sequence Hypothesis, says that:

> the specificity of a piece of nucleic acid is expressed solely by the sequence of its bases, and that this sequence is a (simple) code for the amino acid sequence of a particular protein. (Crick 1958, 152)

In other words, the genetic information in a DNA molecule consists of some discrete set of nitrogen bases in the molecule. The sequence ATC might translate into one instruction, the sequence TTC into another instruction, the sequence CGC into yet another instruction, and so on. The most likely—but, again, not known—likelihood was that base sequences carry information about the synthesis of amino acids. Amino acids are the basic units from which proteins are made, and the primary function of DNA is to tell a cell how to make proteins.

At this point, Crick (and all other scientists) knew virtually nothing about the code itself, although one piece of information seemed to be obvious: the number of nitrogen bases needed for a specific instruction. Suppose a single nitrogen base codes (carries the information needed) for the synthesis of some specific amino acid, as, for example, adenine codes for the amino acid aspartic acid. But that code cannot work because there are only four nitrogen bases and more than 20 amino acids needed for the synthesis of proteins. Similarly, a two-base code cannot work because the greatest number of amino acids coded for with a two-base system is 16 (4 × 4). But a three-base code would provide enough combinations to code for all known amino acids found in proteins.

Crick called his second hypothesis the Central Dogma. In its simplest form, this hypothesis says that once "information" has passed into protein it cannot get out again. That is, nucleic acids are able to make other nucleic acids or proteins, but proteins are never able to make nucleic acids.

A number of scientists misunderstood Crick's argument here. They thought that he was saying that the information stored in DNA molecules may be transferred to RNA molecules, which are then used to synthesize proteins. And this information pathway is certainly correct. We now know that it is the primary mechanism by which the information stored in DNA molecules directs the synthesis of proteins. But, as it turns out, other pathways are also possible. For example, RNA molecules are sometimes used to make new DNA molecules. But, as the Central Dogma says, the one prohibited transfer is from a protein molecule to any other kind of molecule, protein, or nucleic acid.

The obvious next step in the refinement of DNA technologies was to discover the genetic code, the set of three nitrogen bases (sometimes called a *triad* or a *codon*) that codes for amino acids. That breakthrough occurred in 1966 when American biochemists Marshall Nirenberg and Robert Holley discovered the first element in the code. In an elegantly simple experiment, they prepared a sample of RNA made of only one nitrogen base, uracil. They called the molecule polyuracil RNA because it contained many (poly-) uracil units joined to each other. When polyuracil RNA was inserted into a cell whose own DNA had been removed, the cell made only one product, the amino acid phenylalanine. The codon UUU, therefore, codes for the amino acid phenylalanine. The first clue to the genetic code had been produced.

The Nirenberg-Holley experiment not only provided the first "letter" in the genetic code, UUU = phenylalanine, but also suggested a method for breaking the rest of the code. Other researchers were soon constructing synthetic RNA molecules with known base sequences and learning which sequence was responsible for the synthesis of which amino acid. Within a short time, the complete genetic code was known. That code is shown in Figure 1.6.

		2nd base			
		T	C	A	G
1st base	T	TTT = Phe TTC = Phe TTA = Leu TTG = Leu	TCT = Ser TCC = Ser TCA = Ser TCG = Ser	TAT = Tyr TAC = Tyr TAA = STOP TAG = STOP	TGT = Cys TGC = Cys TGA = STOP TGG = Trp
	C	CTT = Leu CTC = Leu CTA = Leu CTG = Leu	CCT = Pro CCC = Pro CCA = Pro CCG = Pro	CAT = His CAC = His CAA = Gln CAG = Gln	CGT = Arg CGC = Arg CGA = Arg CGG = Arg
	A	ATT = Ile ATC = Ile ATA = Ile ATG = Met/START	ACT = Thr ACC = Thr ACA = Thr ACG = Thr	AAT = Asn AAC = Asn AAA = Lys AAG = Lys	AGT = Ser AGC = Ser AGA = Arg AGG = Arg
	G	GTT = Val GTC = Val GTA = Val GTG = Val	GCT = Ala GCC = Ala GCA = Ala GCG = Ala	GAT = Asp GAC = Asp GAA = Glu GAG = Glu	GGT = Gly GGC = Gly GGA = Gly GGG = Gly

Figure 1.6   The Genetic Code

Recombinant DNA Technology

By the mid-1960s, scientists had developed a reasonably satisfactory model of the way DNA molecules reproduce themselves and direct the synthesis of proteins in cells. They were beginning to imagine ways in which they could manipulate that process artificially. In order to do so, they had to develop tools and procedures for working with individual DNA molecules in order to make the transformations they desired. One of the first steps in that direction arose from the work of the Swiss microbiologist Werner Arber and his colleagues. Arber was interested in the fact that bacteria appear to have evolved a mechanism for protecting themselves against attack by viruses (called *bacteriophages*, or just *phages*). That mechanism involves the use of enzymes with the ability to make scissor-like cuts in the DNA of the invading phage particles. Werner and his team were able to isolate the specific enzymes used by bacteria for this purpose, enzymes that were given the name *restriction endonuclease* or *restriction enzyme*.

This discovery, for which Arber received a share of the 1978 Nobel Prize for Physiology or Medicine, provided researchers with the first tool they needed in working with DNA molecules, a way of slicing open the molecule. The additional feature needed for that tool, however, was a "recognition" feature that could be used to cut a DNA molecule at any desired and specific point, such as the bond between an adenine nucleotide and a cytosine nucleotide ($A - C \rightarrow A + C$), or between a guanine nucleotide and thymine nucleotide ($G - T \rightarrow G + T$). That step was accomplished in 1970, when American molecular biologist Hamilton O. Smith and his colleagues discovered a restriction enzyme that recognizes a specific nitrogen base sequence in a DNA molecule. That enzyme, which they called *endonuclease R*, but is now known as *HindII*, "recognizes" the base sequence shown here and cuts it in the center of the sequence, as shown by the arrows.

More than 3,000 restriction enzymes have now been discovered, each with the unique property of recognizing and cutting a DNA molecule within some specific base sequence. Many of these enzymes are commercially available for "off-the-shelf" use by researchers. With these enzymes, researchers now have a way of cutting apart a DNA molecule at virtually any point within its structure.

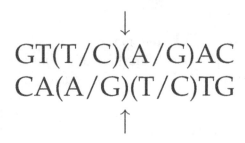

$$\downarrow$$

$$GT(T/C)(A/G)AC$$
$$CA(A/G)(T/C)TG$$

$$\uparrow$$

A "cutting tool," such as restriction enzymes, is essential in manipulating a DNA molecule since it provides a way of opening up the molecule. But another kind of tool is also necessary, one that seals the molecule once a desired change has been made in it. By the mid-1960s, scientists already knew that such tools must exist in nature. That knowledge comes from the fact that DNA molecules have the ability to repair themselves after being damaged. For example, DNA that has been damaged by exposure to X-rays is often found later to have been repaired by some mechanism. That mechanism, researchers decided, must be some kind of enzyme that "knits up" the broken pieces of DNA and makes it functional again. The search was on for that enzyme.

In 1967, a molecular geneticist at the National Institutes of Health, Martin Gellert, reported that his research team had found the putative compound, an enzyme known as a *ligase* (from the Latin *ligare*, "to bind"). More specifically, since this enzyme repairs damage to DNA molecules by restoring bonds that have been broken, the compound is known as *DNA ligase*. The discovery of ligases was of significance, because researchers now had the tools both to cut DNA molecules (restriction enzymes) and to put the molecules back together again (DNA ligases).

The first successful use of these tools in the manipulation of a DNA molecule was accomplished in 1972 by a research team headed by American biochemist Paul Berg. Berg worked with two well-studied viruses, the SV40 (for "simian virus 40") monkey virus and a bacterial virus known as the λ (lambda) bacteriophage. The DNA in both viruses consists of closed loops. The first step in this experiment, as shown in Figure 1.7, involved cutting open the phage DNA with a restriction enzyme, converting it into a linear molecule. A section of the phage DNA was then removed and treated at both ends with chemical groups that made it "sticky"; think of a small strip of Velcro at each end of the strip. Another restriction enzyme was used to cut open the SV40 viral loop, and Velcro-like groups were also added to the ends of the open loop. The "sticky" strip of phage DNA was then attached to opposite ends of the SV40 open chain, and the SV40-λ combination was sealed up by treatment with DNA ligase and a

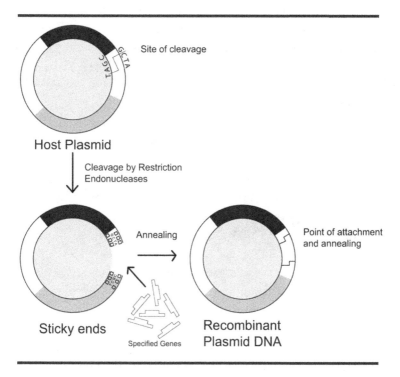

Figure 1.7   Making Recombinant DNA

variety of other enzymes. The product of this experiment was a single molecule of DNA consisting of base sequences from two different organisms. It was a *recombinant DNA* (rDNA) molecule, the first ever made by humans. The molecule could also be described in other ways, as a chimera, for example, or as a transgenic molecule. The word *chimera* comes from the Greek mythological animal with the body and head of a lion, the tail of a snake, and a goat's head protruding from its spine. The term *transgenic*, similarly, refers to any organism or material resulting from one or more genes from a foreign organism being transplanted into a host organism.

The next step beyond Berg's research was already under way by the time his results were published. At Stanford University, medical researcher Stanley N. Cohen was studying bacteria that are resistant to antibiotics. He knew that bacteria must have this property because of the action of genes stored on their DNA. In the case of bacteria, this DNA exists in the form of circular loops known as *plasmids*. How genes on the plasmid confer antibiotic resistance to bacteria was his research problem. At virtually the same time, Herbert Boyer at the University of California at San Francisco was working on a line of research similar to that of Berg's, attempting to learn more about the function of restriction enzymes in the common bacterium *Escherichia coli*, usually called simply *E. coli*.

Cohen and Boyer did not know each other and probably had heard little or nothing about each other's work. That situation changed, however, when the two men were both invited to present papers about their work at a joint U.S.-Japan conference on bacterial plasmids held in Honolulu in November 1972. Each was interested to hear about the other's research, and over sandwiches at a local delicatessen, they discussed the possibility of collaborating on a new project. After returning to California, they began a series of experiments that drew on the skills each had developed in working with microorganisms.

Their first experiment was relatively simple in concept, although a technological challenge. They worked with a plasmid that had been invented by Cohen, the pSC101 plasmid (p for "plasmid"; *SC* for "Stanley Cohen"; and *101*, an identifying number). The pSC101 is a very simple plasmid consisting of only two genes. One gene codes for replication of the plasmid, and the second gene codes for resistance to the antibiotic kanamycin. The value of the second gene is that it allows researchers to identify the presence of the gene in an experiment by treating the organisms present with the antibiotic. Using Boyer's expertise, they cut the plasmid with a restriction enzyme known as *Eco RI* and inserted a new gene into the opened plasmid, a gene that codes for resistance to the antibiotic tetracycline. They then used DNA ligase and other enzymes to seal up the plasmid, which now had three genes: one for replication, one for kanamycin resistance, and one for tetracycline resistance. The altered plasmid was then inserted into a colony of *E. coli* bacteria, and kanamycin and tetracycline were added to the colony. Boyer and Cohen found that some bacteria were killed off by kanamycin and some by tetracycline. But some remained healthy, indicating that they had absorbed the altered plasmid into their own genetic material. The experiment demonstrated that it was possible to change the genetic characteristic of an organism (*E. coli* bacteria in this case) by adding a new component (the tetracycline gene in this case) to the host organism.

The next series of Boyer-Cohen experiments was even more impressive. Using the same procedure as described earlier, they added a gene removed from the South African toad (*Xenopus laevis*) to the pSC101 plasmid and, again, inserted the plasmid into a colony of *E. coli* bacteria. After a period of time, they found that the toad gene was expressing itself in the *E. coli* colony, and that it continued to do so generation after generation. Perhaps of equal importance was the consequence of this experiment. Bacteria reproduce quite rapidly; in the case of *E. coli*, reproduction occurs about every 20 minutes. That means that any substance produced by the normal metabolism of the bacteria—the *X. laevis* gene product in this case—could be captured as a product of the bacterium's normal process of reproduction. The bacterium had become a "factory" for the substance coded for by the toad gene (in this case, nothing other than an RNA molecule).

The Boyer–Cohen experiments have become a classic landmark in the history of DNA technology. Cohen chose to remain at Stanford and continue his research on antibiotic resistance in bacteria. Boyer, however, chose a somewhat different future. In 1976, he and venture capitalist Robert A. Swanson founded Genentech, a company whose purpose it was to conduct basic research on recombinant DNA products and to develop those products for commercial use. In that respect, Genentech was the first company founded to commercialize the products of recombinant DNA research. Before it was a year old, the company had developed its first commercial product, genetically engineered somatostatin, a naturally occurring growth hormone inhibiting hormone that inhibits the production of somatotropin, insulin, gastrin, and other hormones.

Boyer was not the only person to recognize the potential commercial value of his discovery with Cohen. At Stanford University, Neils Reimer, then director of the university's newly established Office of Technology Licensing, tried to convince Cohen to file a patent on the Boyer–Cohen invention, if it could be called so. At first, Cohen was reluctant to seek financial advantage from a scientific discovery. Eventually he relented, however, and three patent applications were filed in 1974, one for the methodology used in the research and two for any products that might result from use of the procedure (one for prokaryotes and one for eukaryotes). The applications were filed in the name of the university. The U.S. Patent Office agonized for a long time over the question as to whether living organisms and procedures related to them could be patented, but eventually decided to issue the patents, which it did on December 2, 1980, August 28, 1984, and April 26, 1988. The three patents all expired in 1997. By the end of 2001, Stanford and the University of California had realized more than $255 million in profits from licenses to 468 companies. A significant amount of that profit came from 2,442 products that had been developed from the patented procedures. Licensing fees were a modest cost to the companies who had to pay them, considering that the 2,442 products had generated an estimated $35 billion is sales over the life of the patents (Beera 2009, 760–761).

Cloning

An important consequence of the Boyer–Cohen experiments was their possible adaptation for cloning. Cloning is the process by which an exact copy is made of a gene, a cell, or an organism. Some of the earliest research on cloning dates to the studies of the German embryologist Hans Spemann in the 1920s; he developed fundamental methods for cloning cells and organisms. Later, researchers found ways to clone tadpoles (Robert Briggs and Thomas King, 1952), carrots (F. E. Steward, 1958), frogs (John Gurdon, 1962), and mice (Karl Illmensee, 1977). But Boyer and Cohen provided a method for cloning at the most fundamental level. When a specific segment of DNA is inserted into a bacterium and the bacterium begins to reproduce, it coincidentally makes exact copies of the original DNA segment. After many generations, the bacterium will have produced dozens, hundreds, or thousands of exact copies—clones—of the original DNA segment.

Over the past three decades, methods of cloning genes, cells, and whole organisms have improved significantly, resulting in the production of cloned sheep, mice, monkeys, cows, cats, deer, dogs, horses, mules, oxen, rabbits, and rats (National Human Genome Research Institute 2009). The technology has also been used in an attempt to help save endangered species and, remarkably, to produce clones of species that have already gone extinct, raising at least the theoretical possibility that those species might once more be returned to life on Earth. (See, for example, De-Extinction 2016; Quill 2015.)

Transgenic Plants and Animals

The work of Berg, Boyer, Cohen, and their colleagues demonstrated the feasibility of creating transgenic organisms, organisms with DNA from two or more different organisms. During the late 1970s and 1980s, researchers carried out a number of experiments designed to confirm this prediction and to create ever more complex transgenic bacteria, plants, and animals. An example of this research was the work of the German American biochemist Rudolf Jaenisch. In the early 1970s, Jaenisch discovered a method for inserting genes in the early embryo of a mouse. When that embryo developed and became an adult, its own DNA had incorporated the foreign genes injected during the embryonic state. A 1974 paper by Jaenisch reported on the production of this research, the production of the first genetically engineered animal (Jaenisch and Mintz 1974). Over the next three decades, researchers produced an almost bewildering variety of transgenic animals, including transgenic pigs, sheep, dairy cattle, chickens, fish (such as the GloFish®), and laboratory animals, such as mice and rats (for a review of these developments, see Lievens et al. 2015; Melo et al. 2007).

Research on transgenic plants was also moving forward at the same time. In a somewhat unusual event, the first transgenic plants were introduced by four different research groups within a few months of each other. Three of those groups presented papers about their discoveries at a conference in Miami, Florida, in January 1983, while the fourth group announced its own discovery at a conference in Los Angeles in April of the same year. The three groups who reported in January had all used similar approaches for the insertion of a gene providing resistance to the antibiotic kanamycin in tobacco plants, in two cases, and petunia plants, in the third case (Fraley, Rogers, and Horsch 1983; Framond 1983; Schell et al. 1983). The fourth group took a somewhat different approach and introduced a gene removed from the common bean plant and inserted it into a sunflower plant (Murai et al. 1983).

One of the interesting questions raised by research on transgenic plants and animals was whether life forms could be patented. When Stanford submitted applications for the procedures and products developed by Boyer and Cohen in 1974, the U.S. Patent Office (now the U.S. Patent and Trademark Office; USPTO) was uncertain as to how it should respond. It had never before been asked to patent living organisms, the products of their metabolism, or the procedures by which they functioned. As noted earlier, it waited six years before making its first decision about this question. One issue involved in its decision was how the courts would act on applications of this kind. That roadblock was removed in 1980 when the U.S. Supreme Court ruled that the Patent Office could issue a patent to the General Electric Company for an organism invented by the Indian American chemist Ananda Chakrabarty. Chakrabarty had used recombinant DNA technology to develop a bacterium from the *Pseudomonas* genus that was capable of breaking down crude oil. He anticipated that the bacterium would be useful in cleaning up oil spills. At first, the Patent Office rejected Chakrabarty's application for a patent because it did not think that living organisms could be patented. Chakrabarty and General Electric appealed the Patent Office's decision until the case reached the Supreme Court. On June 16, 1980, Chief Justice Warren Burger wrote, "A live, human-made micro-organism is patentable subject matter under [section] 101 [of Title 35 of the U.S. Code]. Respondent's micro-organism constitutes a 'manufacture' or 'composition of matter' within that statute" (*Diamond v. Chakrabarty* 1980, 446). In reaching this decision, Burger (and the Court majority) reviewed the position of the U.S. Congress on the issuance of patents for plants in its discussions prior to adopting the Plant Patent Act of 1930. At that time, the position of the Congress was that

> There is a clear and logical distinction between the discovery of a new variety of plant and of certain inanimate things, such, for example, as a new and useful natural mineral. The mineral is created wholly by nature unassisted by man. . . . On the other hand, a plant discovery resulting from cultivation is unique, isolated, and is not

repeated by nature, nor can it be reproduced by nature un-
aided by man . . . (Senate Report No. 315 and House Report
No. 1129, as cited in *Diamond v. Chakrabarty* 1980, 447)

From these statements, Burger concluded that

Congress thus recognized that the relevant distinction was
not between living and inanimate things, but between
products of nature, whether living or not, and human-made
inventions. (*Diamond v. Chakrabarty* 1980, 447)

The Court's decision was of profound significance, because it
meant that a researcher could apply for a patent on virtually any
novel living organism, the method by which it was produced,
the tools and techniques developed for its production, and any
products that might be obtained from the organism. The first
beneficiaries of the Court's 1980 decision were researchers working with bacteria, viruses, and other lower forms of life. Before
long, however, researchers working with organisms at every
level—all kinds of plants and animals—had begun to seek patents for their inventions and discoveries. The DeKalb Genetics
company, for example, sought a patent for a genetically engineered corn plant with a greater than average level of the amino
acid tryptophan, improving its nutritional value for animals to
whom it was fed. The USPTO rejected DeKalb's request, a decision that was later overruled by the Board of Patent Appeals. The
patent was issued in 1990, the first patent for a transgenic plant.

At about the same time, researchers at Harvard University
were developing the oncomouse, a mouse that has been engineered to carry an additional gene that places the animal at
significantly greater risk for cancer than its non-engineered
cousins. The value of the mouse is that it can be used in cancer research. The USPTO issued a patent for the oncomouse
on April 12, 1988, and its name was registered as the Onco-
Mouse® by the DuPont company, which holds licensing rights
to the invention. The inventors also applied for a patent in
Europe in June 1985, and after a very long series of denials

and appeals, the patent was granted on July 6, 2004. Only in Canada has the application process failed. On December 5, 2002, the Canadian Supreme Court ruled that the mouse was not an "invention" and that it could not, therefore, be patented in Canada ("Bioethics and Patent Law: The Case of the Oncomouse" 2006).

The legality of patenting life forms continued, understandably, to be a matter of dispute. From time to time, cases arose that questioned the right of a company to patent a plant, an animal, a protist, a gene, or some other product or methodology associated with recombinant DNA manipulation of living organisms. In 2001, the U.S. Supreme Court again considered this issue. J. E. M. Ag Supply, Inc., doing business as Farm Advantage, Inc., had sued Pioneer Hi-Bred International, Inc., a company that held 17 patents on engineered plants. J. E. M. Ag Supply claimed, the Court's 1980 decision notwithstanding, that genetically engineered plants could not be patented. The case reached the Court on October 3, 2001, and was settled on December 10, 2001. The Court, in a six-to-two ruling, reaffirmed its position, stating that nothing in existing federal law prohibited the patenting of novel organisms, their parts, or technology used to produce them (*J. E. M. Ag Supply, Inc., DBA Farm Advantage, Inc., et al. v. Pioneer Hibred International, Inc.* 2001).

With the legal status of patented life largely settled in the United States, the USPTO has been deluged with applications for patents on almost every conceivable type of recombinant product. Today it is difficult to estimate the number of patents issued by the USPTO in the last three decades in this area. A search of the USPTO database for terms related to such patents granted since 1976 produces very large numbers of patents, more than 3,000 that include the term "organism" in its abstract; more than 20,000 that include the term "gene," more than 23,000 that include the term "animal," and more than 54,000 that mention the word "plant" (USTPO Patent Full-Text and Image Database 2016). It appears that almost every development in the engineering of life forms has become, and will become, the property of some individual or corporation.

Practical Applications of Transgenic Organisms

The work of Berg, Boyer, Cohen, and countless numbers of researchers in the late 1950s, 1960s, and early 1970s provided a bridge between the "what we think can be done" occasioned by the discovery of the DNA structure by Watson, Crick, and Franklin, and the "what CAN be done" of today's DNA technology. The products of today's DNA technology fall primarily into four categories: research, pharmaceuticals, agriculture, and industry.

Research

The OncoMouse®, developed by Harvard researchers Philip Leder and Timothy A. Stewart in the late 1980s, is an early example of the way a genetically modified organism can be used in research. Since the OncoMouse® is engineered to contain an oncogene, a gene that tends to cause cancer in an organism, the engineered organism provides an ideal system for the study of cancer and methods that can be used for its treatment. A somewhat different example is the knock-out mouse, first created by Italian American molecular geneticist Mario R. Capecchi, British developmental biologist Martin Evans, and British American geneticist Oliver Smithies in 1989 (for which they won the 2007 Nobel Prize in Physiology or Medicine). The knockout mouse is a mouse that has been genetically modified by the inactivation of one or more genes. This change allows a researcher to determine precisely the function of some particular gene that has been "knocked out."

Pharmaceuticals

One of the most productive uses of recombinant DNA technology has been the use of bacteria for the production of pharmaceuticals. The procedure used in this field is identical to that employed by Berg, Boyer, and Cohen. First, a gene coding for the production of some desired protein is obtained, either by removing it from an organism or by synthesizing it in the laboratory. That gene is then inserted into a plasmid (as in the Berg and Boyer–Cohen experiments), a virus, or some other structure, called a *vector*. The vector is then inserted into a host cell, most commonly a bacterium. If all goes well, the bacterium incorporates the engineered plasmid into its genetic material and begins to produce the protein coded by the inserted gene along with its normal metabolic products. When the bacterium reproduces, it produces two copies of itself (then four, then eight, then 16, and so on), each of which continues to produce the desired protein. A large vat of bacteria becomes, then, a microbial factory for the generation of the desired protein product.

This method has now been used to produce hundreds of different therapeutic proteins, more than 130 of which have been approved for commercial use by the U.S. Food and Drug Administration (FDA). Table 1.1 summarizes some of the drugs that have been prepared by rDNA technology and have been approved for use by the FDA (for a review of the current status of the manufacture of drugs by genetically modified organisms, see Leader, Baca, and Golan 2008; "Alphabetical List of Licensed Products" 2016).

Bacteria provide a simple and inexpensive mechanism for the production of recombinant pharmaceuticals. But they are by no means the only way of making drugs with the use of rDNA technology. For nearly two decades, researchers have been considering the use of farm animals for the production of useful proteins. Animals have a number of advantages over bacteria for the synthesis of human proteins. They have a built-in system for making proteins that operates at the correct temperature and pH (acidity), and a natural immune system provides an ideal environment for the culturing and harvesting of drugs produced from transplanted genes. They

Table 1.1   Some Genetically Engineered Drugs Approved by the FDA

Drug	Year Approved[1]	Approved Use
Human insulin	1982	Diabetes mellitus
Human growth hormone	1985	Growth deficiency in children
Alpha interferon (2b)	1986	Hairy cell leukemia; genital wards, Kaposi's sarcoma, hepatitis B; hepatitis C
Hepatitis B vaccine	1986	Hepatitis B
Alterplase	1987	Acute myocardial infarction; acute massive pulmonary embolism
Epoetin alfa (erythropoietin)	1989	Anemia associated with chronic renal failure
Gamma interferon (1b)	1990	Chronic granulomatous disease; severe, malignant osteopetrosis
Antihemophiliac factor	1992	Hemophilia A
Proleukin (IL-2)	1992	Metastatic kidney cancer; metastatic melanoma
Beta interferon (1b)	1993	Certain types of multiple sclerosis
Imiglucerase	1994	Gaucher's disease
Pegaspargase	1994	Acute lymphoblastic leukemia
Coagulation factor IX	1997	Factor IX deficiencies
Glucagon	1998	Hypoglycemia in diabetics
Adalimumab	2003	Rheumatoid arthritis
Platelet-derived growth factor-BB	2005	Periodontal bone defects and associated gingival recession
Hyaluronidase	2005	As an adjunctive to other injected drugs
HPV vaccine	2006	Human papillomavirus disease
Sorafenib	2007	Primary kidney cancer; primary liver cancer
Ipiliumab	2011	Unresectable or metastatic melanoma
Pertuzumab	2012	HER2-positive metastatic breast cancer
Eleyso	2012	Gaucher's disease
Kovaltry	2016	Antihemophilic factor in children and adults

[1] Additional approvals may be given for similar or related products at a later date.

also have all the nutrients needed to keep the system operating efficiently, and they possess a natural output system in the form of urine, blood, and/or milk from which manufactured drugs can be collected.

One of the pioneers in the field was PPL Therapeutics, the company that was responsible in 1996 for the development of the first cloned mammal, the sheep known as Dolly. In 1990, PPL transplanted a gene for the production of the protein alpha-1-antitrypsin (AAT) into the DNA of a sheep named Tracy. AAT is normally produced in the liver and transported to the lungs, where it is used to support breathing. Individuals who lack the protein are short of breath and have trouble breathing when carrying out even the simplest activities.

The experiment appeared to be very successful. More than half of the protein in milk produced by Tracy was AAT. And her descendants were equally productive. By 1998, PPL reported that Tracy had more than 800 granddaughters who were producing an average of about 15 grams per liter of AAT per day. That success was short-lived, however, as clinical trials with humans using the engineered protein found that a number of patients developed an unexplained, but troublesome, "wheezing" problem. PPL decided to discontinue clinical trials of the engineered AAT, and the trials have not since been restarted (Colman 1999).

The problems experienced with Tracy have not discouraged other researchers or pharmaceutical corporations from pursuing the use of farm animals for the production of drugs, a field now commonly known as *pharming*, for "pharmaceutical development using farm animals." This technology is also being used with plants, a technology described as *plant molecular farming* (PMF), or, simply, *molecular farming*. Researchers are currently exploring a number of possible uses for PMF in the production of useful proteins, such as the protein gastric lipase for the treatment of cystic fibrosis in corn plants (Meristem), antibodies for the treatment of the common cold and dental caries in tobacco plants (Plant Biotechnology), intrinsic factor for the treatment of vitamin B12 deficiency in arabidosis plants (Cobento), lactoferrin for the prevention of infections in corn (Meristem), and edible oral vaccines in potatoes and tomatoes (Arizona State University) ("Pharming the Future" 2016). Some of the pharmed drugs currently under development are listed in Table 1.2.

Table 1.2   Some Pharmed Products Currently under Development

Pharmaceutical	Host Animal	Target Disorder(s)
Alpha-1-antitrypsin	Sheep	Emphysema; cystic fibrosis, other lung disorders
Cystic fibrosis transmembrane conductance regulator (CFTR)	Sheep	Cystic fibrosis
Tissue plasminogen activator	Sheep, pig	Thrombosis
Factor VIII and factor IX	Sheep, pig, cows	Hemophilia
Fibrinogen	Sheep, cows	Wound healing
Human protein C	Goat	Thrombosis
Collagen I and collagen II	Cow	Anti-arthritic; tissue repair
Human serum albumin	Cow, goat	Surgery, burns, shock
Human fertility hormones	Cow, goat	Human infertility
Lactoferrin	Cow, maize	Gastrointestinal tract disorders, anti-arthritic
Erythropoietin	Rabbit	Anemia associated with dialysis
Interleukin-2	Rabbit	Renal cell carcinoma
Human growth hormone	Rat	Pituitary dwarfism
Gastric lipase	Maize	Cystic fibrosis; pancreatitis
Hepatitis B vaccine	Lettuce, potato	Hepatitis B
Norwalk virus vaccine	Potato	Norwalk virus disease
Insulin	Safflower	Diabetes
Cytokine (transforming growth factor-beta 2)	Tobacco	Ovarian cancer

Source: Some of the information in this table comes from *Pharmaceutical Biotechnology*, by Daan J. A. Crommelin, Robert D. Sinclair, and Bernd Meibohm, 154.

An important breakthrough occurred in 2006 when the FDA gave approval for the use of the first product from a genetically altered mammal. The product is Atryn®, a recombinant form of human antithrombin III, a small protein that prevents excessive clotting of blood. The product is made by the GTC Biotherapeutics company (Hall 2009).

Agriculture

As noted earlier, humans have for centuries used selective breeding to improve the quality of the crops and animals they grow. They have tried to develop cows that produce more milk, sheep with better wool, stronger horses, corn that will be more resistant to disease, and larger, more nutritious fish. DNA technology has made it possible for plant and dairy farmers to achieve these same objectives more rationally and more efficiently. For example, PPL Therapeutics announced in February 1997 that it had developed a transgenic cow named Rosie whose milk contained the human protein alpha-lactalbumin. The protein was coded for by a gene inserted into Rosie's own DNA when she was still at the embryonic stage. The presence of the human protein in Rosie's milk made it nutritionally superior for humans than is natural cow's milk (Begley 1997). A decade later, researchers were still refining the technology for recombinant produced milk with these qualities (Lu et al. 2016).

Some of the characteristics for which transgenic animals are produced are the following:

- Increased rate of growth;
- Greater muscle mass and/or bone strength;
- Greater resistance to pain;
- Leaner, more tender pork and beef;
- Resistance to disease;
- Increased production of milk and/or for longer periods of time;

- Improved nutrition of eggs (for chickens);
- Milk that lacks human allergens;
- Milk that produces increased amounts of cheese and yogurt;
- Modifications in physical conditions that increase the efficiency of farming, such as the production of hornless cattle;
- Improved feeding of stock, as in increasing the ability of animals to digest cellulose;
- Fish that grow faster than conventional fish;
- Fish that reach sexual maturity at a younger age than conventional fish;
- Fish that tolerate a wider variety of environmental conditions, such as colder or warmer water.

One example of an apparent success in this line of research was the 2005 announcement of the development of a mastitis-resistant cow developed by researchers at the U.S. Department of Agriculture (USDA). Mastitis is an infection of the udder, a common and potentially serious disease among dairy cows, significantly affecting their milk production. Vaccines and antibiotics are currently available for the prevention and treatment of mastitis, but they are not very effective. In the USDA experiment, researchers constructed a gene that codes for the protein lysostaphin, which kills the bacterium that causes mastitis. They then inserted the gene into the DNA of a number of cow embryos. When those cows grew to maturity, their milk contained the lysostaphin protein, and they had a dramatically lower rate of mastitis infections than did a group of control cows (Fiala 2005; Liu et al. 2014).

The development of transgenic animals for agricultural applications is still at the development stage. One reason that progress has been slow is that the addition of a foreign gene to an animal's natural genome sometimes has unexpected and serious side effects. Perhaps the most frequently cited case of such side effects is the so-called Beltsville pigs from the late 1980s. A group of pig embryos received human growth hormone in an attempt to increase their rate of growth and reduce body fat. The experiment appeared to be at least partially successful as the mature pigs' body fat was reduced and their feeding efficiency did improve. However, 17 of the 19 pigs who had received the gene died in the first year, and all animals exhibited a number of debilitating disorders, including diarrhea, mammary development in males, lethargy, arthritis, lameness, skin and eye problems; loss of libido; and disruption of estrous cycles. Because of these problems, the experiment was discontinued (Committee on Defining Science-Based Concerns Associated with Products of Animal Biotechnology. Committee on Agricultural Biotechnology, Health, and the Environment. National Research Council 2002, 98).

A somewhat different use of transgenic animals is in *xenotransplantation*, the transplantation of an organ or other body part from one species to a different species. Over the past 30 years, a number of experiments have been conducted involving the transplantation of hearts, lungs, kidneys, livers, and cells from mammals, such as rabbits, pigs, goats, sheep, and various non-human primates. The most common problem with this type of procedure is that the human host body usually rejects the transplanted organ. The body's immune system recognizes the pig, sheep, cow, baboon, or other organ as a foreign object and mounts its full immune arsenal to destroy the invader, almost always with success. One possible solution to this problem is to engineer the donor animal's DNA so that it more closely resembles that of a human. In an example of this type, Jeffrey Platt at Duke University has explored the possibility of using genetically engineered pig livers as "bridge" (temporary replacement) organs for patients with life-threatening renal failures waiting for a human transplant (Frederickson 1997, 439). Platt reported in 2000 that two patients who had received pig livers transformed to include human genes designed to reduce rejection had survived for periods of five and 18 months (Levy et al. 2002).

DNA technology has also made it possible to modify plants, so as to introduce a number of desirable properties: resistance to disease, protection against attack by insects, improved nutritional characteristics, and reduced rates of spoilage, for example. One of the first genetically modified (GM) agricultural products to be developed was a biological control agent called Frostban. The agent was developed as a way of protecting crops from freezing. Scientists have long known that crops will not freeze even when the temperature drops below 0°F in the absence of certain frost-promoting bacteria. Frostban was a genetically altered form of the common *Pseudomonas* bacterium that attacked and destroyed those bacteria. Developed and field-tested in the early 1980s, Frostban never reached the marketplace because of some serious problems with late field tests and approval issues from the federal government (Baskin 2011).

Another early example of failed GM products was the Flavr Savr tomato developed by Calgene, Inc., in the early 1990s. The tomato contained a gene that interferes with the production of the enzyme polygalacturonase, which causes a tomato to soften. Lacking this enzyme, a tomato could remain on the vine and ripen before being picked, in contrast to conventional tomatoes, which are usually picked while they are still green. The Flavr Savr tomato performed well in field tests, and Calgene requested approval from the FDA in 1991 for its commercial sale. In May 1994, the FDA granted its approval, noting that the Flavr Savr was "as safe as tomatoes bred by conventional means" (Stone 2004). Calgene began marketing the Flavr Savr tomato in 1994 but had little success with the product. Critics have pointed to a number of reasons for its failure in the marketplace, including the reluctance of consumers to purchase genetically modified foods and a poorly conceived marketing plan. In any case, it was withdrawn from the market shortly after Calgene was purchased by Monsanto in 1995 (Huffman 2010).

Since the 1990s, researchers have experienced far more success in the development of genetically engineered crops with a range of desirable properties. These products tend to fall into three categories outlined next.

GM Products Resistant to Certain Pests

Some examples of these products include GM corn that is resistant to the European corn borer, stalk borer, southwestern corn borer, armyworm, and corn earworm; GM cotton that is resistant to the cotton bollworm, the pink bollworm, and the tobacco budworm; and GM potato that is resistant to the Colorado potato beetle. These products are sometimes scientifically successful, but commercially a failure. For example, the GM potato described here, marketed under the brand name of New-Leaf, was approved for sale by the FDA, but was withdrawn from the marketplace in 2001 because of poor sales.

GM Products Resistant to Herbicides

Herbicides are widely used by farmers to control weeds that compete with their crops. One of the best known and most widely used herbicides is Roundup®, manufactured by Monsanto. The most serious problem with such herbicides is that they often kill crops along with the weeds. The solution to this problem has been to develop, by DNA technology, crops with foreign genes that protect them against herbicides. One of the most successful of these new products is so-called Roundup Ready® crops, crops that are not harmed when sprayed with Roundup®. Today, seeds for a number of Roundup Ready® crops are available from Monsanto, including those for cotton, corn, canola, lettuce, sugar beets, soybeans, tobacco, tomato, and

wheat. GM crops resistant to herbicides other than Roundup®
are also under development or in use for these and other crops,
including alfalfa, barley, flax, melons, peanuts, potatoes, rice,
sweetgum, tobacco, and tomato (Fernandez-Cornejo, Jorge,
et al. 2014).

GM Products with Nutritional Value or Other Benefits

In the 1990s, Swiss botanist Ingo Potrykus began research on
a genetically modified form of rice capable of synthesizing beta
carotene, a precursor of vitamin A. Potrykus was interested in
this problem because vitamin A deficiency is responsible for
an estimated 250,000–500,000 cases of blindness in children
around the world. If he could improve the nutritional quality
of the primary food for the vast majority of those children,
he reasoned, he could significantly reduce this serious public
health problem. The product he invented, called golden rice, is
a form of conventional white rice that includes three inserted
genes that code for beta carotene. It is the first genetically engineered product made specifically to improve the nutritional
value of a natural food. The method Potrykus used, however, is
applicable in principle to a comparable treatment of virtually
any natural food.

One of the increasingly serious problems faced by farmers
around the world is insufficient land area on which to grow
their crops. More and more, they are using land that lacks sufficient water, has too high a salt content, or is unsatisfactory
for some other reason. For over a decade, researchers have been
looking for ways to genetically modified crops to be more tolerant of drought and high salt concentrations. One approach has
been to insert the gene for the enzyme H^+-pyrophosphatase,
which appears to make plants more resistant to drought and
to sodium chloride (NaCl). In one series of experiments along
this line, engineered plants with the additional genes for
H^+-pyrophosphatase were able to survive longer with water
and better when exposed to high concentrations of salt than
were conventional plants (Gaxiola et al. 2001, 11444–11449).

Research on drought and salt-tolerant plants is well under way in a number of laboratories around the world, although it is still not clear what commercial value these plants may have in the near future (Waltz 2014).

Industry

In 2002, researchers from Nexia Biotechnologies Inc. and the U.S. Army Soldier and Biological Chemical Command (ASBCC) announced that they had manufactured silk fibers normally produced by spiders using recombinant DNA technologies. That discovery was important because, while spider silk is generally recognized as the strongest fiber in the world, no commercial system for extracting the material from spiders has ever been developed. The Nexia and ASBCC researchers had achieved this goal by inserting spider genes controlling the synthesis of silk into goats. Milk produced by the goats contains the spider silk protein, which can then be separated, purified, and spun into threads. This GM material, called BioSteel®, is still under research, but could be used in the manufacture of implants, medical products, textile products, or bulletproof material.

Industrial applications of DNA technology have not developed as rapidly as have those in pharmaceuticals and agriculture, but they hold as much potential. Just a few of the myriad possible products that may result from this line of research are:

- Modifying fats and oils produced by plants to obtain products for use in paints and manufacturing processes;
- Production of polymeric materials from corn plants for the production of plastics;
- Modifying the genetic structure of plants to make them sensitive to certain external stimuli, making them useful as biosensors;
- Engineering of plants to make possible the synthesis of artificial forms of natural and synthetic fibers, such as cotton, wool, and nylon;

- Modifying flowering plants to give them more intense and/ or more interesting flower colors and odors;
- Production of enzymes that can be used to convert biomass into fuels more efficiently than current methods.

Gene Therapy and Genetic Testing

Two other fields of medicine that have benefited significantly from the development of DNA technology have been gene therapy and genetic testing. Gene therapy (also known as human gene therapy, or HGT) is a procedure by which genes are inserted into the body of a person with a genetic disorder. Researchers first began to appreciate the potential usefulness of gene therapy in the 1970s when recombinant DNA procedures were first being developed. Gene therapy involves five basic steps:

1. The gene responsible for a genetic disorder must be identified. A genetic disorder develops because a patient lacks an essential gene or has a faulty copy of that gene, such that some essential protein is either not made or is made incorrectly. For example, the disorder known as phenylketonuria (PKA) occurs because a person lacks the gene needed to make the enzyme hepatic phenylalanine hydroxylase. Without this enzyme, the body cannot metabolize the amino acid phenylketonuria, and a disease develops.

2. A correct copy of that gene must be obtained from some natural source or must be produced synthetically in the laboratory.

3. That gene must be introduced into the patient's DNA. rDNA procedures for splicing a gene into a DNA molecule developed in the 1970s provided a mechanism for inserting a correct gene into a patient's DNA.

4. The altered DNA must then be introduced into cells from the patient's body. Originally, this step was performed by removing a sample of blood from the patient's body and inserting corrected DNA into lymphocytes (white blood

cells) in the blood. A virus was typically used as a vector (carrier) for the inserted gene, although alternative vectors and methods of insertion are now available.

5. Cells containing the altered (correct) DNA are then returned to the patient's body where, if the procedure is successful, they begin to synthesize the missing protein. They must also continue to do so as the lymphocytes (or other engineered cells) reproduce over and over again in the patient's body.

After more than five years of preparation, the first gene therapy trial was conducted on September 14, 1990, by a group of researchers led by American medical researcher and molecular biologist W. French Anderson. Their subject was a four-year-old girl with adenosine deaminase (ADA) deficiency, a condition that leaves a person susceptible to all kinds of infections. The condition is usually fatal at an early age. Anderson's team followed the general procedure described earlier, providing the patient with the gene needed to synthesize the adenosine deaminase enzyme she was lacking. Although the procedure had to be repeated a number of times following the original experiment, it appeared to have been at least partially successful, and the patient remains alive today.

Gene therapy has been plagued by a number of problems and tragedies since the earliest successes of Anderson's research. For example, 18-year-old Jesse Gelsinger died in 1999 while being treated for a rare liver disease called ornithine transcarbamylase disorder. In 2007, 36-year-old Julie Mohr died while being treated for rheumatoid arthritis, although the cause of her death may very well not have been the gene therapy treatment itself. Still, many researchers have great confidence that gene therapy will save the lives of countless humans once the technology used in the procedure is well developed. As of 2015, more than 2,200 gene therapy clinical trials worldwide had been conducted. Table 1.3 shows the number of human gene therapy trials conducted worldwide between 1989 and 2015.

Table 1.3 Human Gene Therapy Clinical Trials Worldwide,
1989–2015

Year	Number of Trials
1989	1
1990	2
1991	8
1992	14
1993	37
1994	38
1995	67
1996	51
1997	82
1998	68
1999	116
2000	96
2001	108
2002	98
2003	85
2004	101
2005	112
2006	117
2007	90
2008	120
2009	81
2010	92
2011	87
2012	101
2013	121
2014	130
2015*	45
Unknown	142

*As of July 2015.

Genetic Testing

Scientists have known for well over a century that certain diseases have a genetic basis. The appearance of hemophilia among the children of Queen Victoria in the mid-19th century was a subject of considerable interest to medical researchers at the time. By the middle of the 20th century, researchers had begun to find an association between specific types of genetic material, such as chromosomes, and specific genetic disorders. In 1959, for example, French geneticist Jerome Lejeune discovered that Down syndrome is caused by the presence of an extra copy of chromosome 21. Still, it took many years for scientists to take advantage of the Watson and Crick discovery and pinpoint specific alterations in DNA molecules that result in genetic disorders.

Efforts to develop methods for testing for genetic disorders date to 1961 when American microbiologist and physician Robert Guthrie developed a test for phenylketonuria (PKU), a genetic disorder in which the body is unable to metabolize the amino acid phenylalanine. The amino acid builds up in the body and may produce serious health effects and, in some cases, death. Once detected, however, the disease is easily treated by providing the patient with a diet low in foods that produce phenylalanine. Guthrie's test was very simple. A drop of blood is drawn from the heel of a newborn child and then tested for the presence of phenylalanine. The amount of the amino acid present in the sample tells whether or not the child has the disorder.

Four kinds of genetic disorders occur. Single-gene disorders are the most common. They occur when a change occurs in a single gene, or the gene is lost entirely. The loss of or damage to the gene means that it is unable to produce some protein essential to a person's well-being. Currently, more than 6,000 single-gene disorders are known. Multifactorial genetic disorders occur when more than one gene is damaged or lost and/ or when an environmental factor is present that moderates the genetic error. Breast cancer, heart disease, diabetes, and hypertension are examples of multifactorial disorders. Chromosomal genetic disorders occur as the result of some change in a whole chromosome, as is the case with Down syndrome. Mitochrondrial genetic disorders are rare conditions that result from errors in the noncoding DNA found in mitochondria. Clearly, the easiest genetic disorders to test for are single-gene disorders, and the vast majority of genetic tests currently available fall into that category.

Until the late 1990s, only gross methods were available for genetic tests. Testing for Down syndrome, for example, involves taking a drop of blood and examining it under a microscope. The condition of chromosomes in the blood can easily be seen, and the presence of trisomy (three copies) on chromosome 21 can be detected. This method works well for any chromosomal genetic disorder. Detecting single-gene disorders is more difficult, however, since one cannot see the specific base sequence present on a chromosome microscopically. The detection of single-gene disorders was made much easier, however, when the complete sequencing of the human genome was completed in the early 2000s.

The Human Genome Project was a program initiated by the U.S. government in October 1990 to map the complete human genome, that is, to determine the exact base sequence for every gene on every human chromosome. That project was completed in 2003, at which point the base sequence for all 20,000–25,000 human genes was known. The value of this research is that scientists now know what the correct base sequence is for every gene in the human genome. Genetic tests can now be devised to determine the base sequence for any given gene from any given person and compare that result with the "normal" base sequence for that gene. If errors are present in the subject's base sequence, it becomes possible, at least in theory, to develop a method to correct that base sequence and to cure that disease. This principle forms the basis of much ongoing research in the development of new genetic tests.

Forensic DNA Testing

The applications of DNA technology thus far can all be characterized as developments in genetic research that have their origins from Mendel's original research in the mid-19th century. Another application, however, was quite unexpected: the use of DNA testing in forensic science. The first breakthrough in this field came in 1984 when British geneticist Alec Jeffreys developed the restriction fragment length polymorphism (RFLP) test for DNA. That test makes use of the fact that a DNA molecule consists of two kinds of base sequences: exons and introns. An exon is a sequence of bases that codes for a protein. Exons tend to be essentially the same from one human (or other animal) to another because all humans need to make the same proteins to function normally. Interspersed between exons are other base sequences that do not code for a protein. In fact, they have no known function and are sometimes known for that reason as "junk DNA." Jeffreys found that introns differ from person to person. Although two individuals are likely to have identical exons, they are likely to have very different introns. He found a method for removing introns from an individual's DNA (using restriction enzymes) and labeling them with radioactive isotopes to permit a radiographic photograph of the removed segments.

Less than a year after this discovery, Jeffreys used the new technology to verify the paternity of a Ghanaian boy who was applying for a passport to visit Great Britain. He compared the DNA "fingerprint" of the boy with that of a man claiming to be his father and found a match, making the boy eligible for entry into the country. A year later, Jeffreys was asked to use his technique in the investigation of a pair of rape-murders in the small town of Narborough in Leicestershire. Jeffreys was able to show that the man being held for the crimes was not guilty, but that some other unknown individual was responsible for both crimes. After further investigation by the police, a new suspect, Colin Pitchfork, was arrested, his DNA was tested by Jeffreys, and a match was obtained. Pitchfork thus became the first person convicted of a crime based on DNA testing (Holden 2014).

Today, Jeffreys's system of DNA analysis by means of RFLP has been largely replaced by a second method of finding, cutting out, and identifying fragments of a DNA molecule. That method was invented by American molecular biologist Kary Mullis in 1986. The technology, called polymerase chain reaction (PCR), is preferable to RFLP, because it takes less time and can be used with much smaller samples of DNA. PCR gets its name from the fact that a very small sample of DNA—in principle, a segment of a single molecule of DNA—can be amplified many times over to produce millions of copies of the original sample, providing a much larger specimen with which to work.

The use of RFLP and PCR in forensic science is based on a statistical principle: DNA taken from two different individuals has only a small likelihood of being exactly the same. Mathematicians have developed models to predict what that likelihood is for various circumstances. In fact, whole books have been written on the subject. As a result, researchers are now able to look at samples taken from a crime scene and from a suspect and say that the chances are one in a million, one in a billion, one in ten billion, or some other number, allowing judges and juries to decide whether a particular individual is likely to have committed a particular crime. Forensic DNA analysis (sometimes called DNA fingerprinting) is now generally acknowledged as being the single most reliable tool for identifying criminal suspects throughout the world.

Recent Advances in DNA Technology

The 21st century has seen the development of a new type of genetic engineering that has revolutionized the field of

DNA technology. This form of genetic engineering is called *gene editing*, and it provides a new mechanism for altering DNA that promises to solve a number of problems in the field that were difficult using traditional methods of recombinant DNA studies.

Gene editing (also known as genome editing) had its origins in the late 1980s with studies of the ways in which prokaryotes protect themselves from attack by agents such as viruses and bacteriophages (phages). Researchers discovered that prokaryotes have evolved a two-part system for fighting off the invasion as such agents. One part consists of a segment of RNA that recognizes a specific segment of DNA in the attacking virus or phage. That recognition capability is similar to systems found in eukaryotic immune system in which an organism "remembers" the molecular structure of an antigen that it has encountered at one point of time and then responds when it encounters the same DNA segment at a later point in time. The second part of the prokaryotic system is a nuclease, a molecule that is capable of cutting the bonds between nucleotides in DNA molecule, thus incapacitating that molecule.

The sequence of events that occurs in a bacterium when attacked by a virus or phase, then, begins when the bacterium recognizes a portion of the viral or phage DNA and synthesizes an RNA segment that is complementary to the DNA segment. The new RNA is then attached to a nuclease capable of attacking the invader DNA. The RNA-nuclease dyad (two-part system) then attacks the invader DNA, attaches at the recognition site, and cuts the DNA molecule, thus inactivating the molecule and the invader's ability to continue functioning within the bacterium.

The first system of this type to be discovered is the so-called zinc-finger nuclease (ZFN) system. The ZFN name comes from the fact that the segment that binds to a DNA segment consists of a so-called *zinc finger*, a small protein molecule that contains a zinc atom at its core. The ability of the ZFN system to break DNA at specified sites in various organisms was discovered in the late 1980s and early 1990s and later adapted for human genomes (Carroll 2011, 774). One disadvantage of ZFN systems is that they recognize only short segments of DNA, usually only a few bases long.

The shortcomings of the ZFN system led to active research for other systems of gene editing that were more efficient and easier and less expensive to use. This search led to the discovery of so-called transcription activator-like effector nuclease (TALEN) systems, which recognize base sequences of up to 30 to 35 bases in length. In 2011, the journal *Nature Methods* recognized ZFN and TALEN as "method of the year" (Baker 2012). A third method of gene editing makes use of even larger molecules known as *meganucleases*, with the more restricted ability to recognize and cut relatively long stretches of DNA. (For a discussion and comparison of these three methods, see Cox, Platt, and Zhang [2015]. For more detailed information about the methods and videos of their operation, see Gaj, Gersbach, and Barbas 2013; "Method of the Year 2011: Gene-Editing Nucleases 2011"; Meganucleases 2010.)

Probably the most popular of gene-editing programs today is the so-called clustered regularly interspaced short palindromic repeats (CRISPR) system. The system was found independently by Japanese researcher Yoshizumi Ishino in 1987 and Spanish researcher Francisco Mojica in 1993. In prokaryotes, the system acts as an immune system by attaching a CRISPR sequence to a short segment of RNA, which then guides the sequence to a matching portion of DNA. When the RNA-CRISPR unit attaches to the DNA, it releases an enzyme, Cas9, which cuts the DNA at a precise location, allowing the removal or insertion of a new segment into the DNA, thus modifying one of the organism's genes. As of 2016, CRISPR has been used for gene editing on a number of different species, including baker's yeast (*S. cerevisiae*), zebra fish (*D. rerio*), fruit flies (*D. melanogaster*), axolotl (*A. mexicanum*), nematodes (*C. elegans*), mice, monkeys, and a variety of plants. It has also been used on the modification of human cells and human embryos. Although the CRISPR technology is still in its earliest phases, it shows quite remarkable promise for the modification of the human genome, a prospect that has become the subject of considerable discussion and controversy over the past few years (About Human Germline Gene Editing 2016; CRISPR-Cas9 and Genome Editing 2016; Genome Editing with CRISPR-Cas9 2014 [video]; Wirt 2013).

Gene editing has long been considered a highly controversial issue, primarily because of its potential use with human beings. The ability to permanently change the fundamental nature as to who and what a person is has generally been regarded as opening the door to a host of social, psychological, medical, and other factors. For that reason, gene editing with human subjects has been banned in a number of countries worldwide. (For a review of national policies on gene editing worldwide, see "How the World's Governments Have Regulated Human Genome Editing" 2016; Isasi, Kleiderman, and Knoppers 2016.)

Conclusion

The study of genetics has come a very long day indeed from the period during which humans attempted to improve the plants and animals in their lives by trial-and-error hybridization. They eventually learned that the changes that occur during this process can be understood and predicted in terms of some "unit" or "factor" present in organisms that behaved by clear mathematical rules. It was not until the research of Watson and Crick in the early 1950s, however, that scientists discovered precisely what this "unit" or "factor" was: a molecule of DNA. In the seven decades since that breakthrough, researchers have developed an amazing number and variety of molecular tools that have allowed them to develop a far better understanding of the structure and function of DNA and its partner, RNA, in species of all kind. Today, researchers are faced with the potential of redesigning all kinds of organisms to perform a variety of functions for the betterment of human kind. They also have at hand the tools with which they can alter the most basic nature of plants and animals—including human beings—in whatever way they choose. With those possibilities come a number of ethical, philosophical, economic, social, political, and theological problems with which humans will have to grapple in the near future. Chapter 2 describes some of those problems and issues and some of the solutions that have been suggested and adopted for them.

References

"About Human Germline Gene Editing." 2016. Center for Genetics and Society. http://www.geneticsandsociety.org/article.php?id=8711. Accessed on April 2, 2016.

"Alphabetical List of Licensed Products." 2016. U.S. Food and Drug Administration. http://www.fda.gov/downloads/biologicsbloodvaccines/ucm149970.pdf. Accessed on March 28, 2016.

Baker, Monya. 2012. "Gene Editing Nucleases." *Nature Methods*. 9(1): 23–26. Available online at http://www.massgeneral.org/pathology/assets/ResearchPath/bernstein_Baker_Nat%20Methods%202011.pdf. Accessed on April 3, 2016.

Baskin, Yvonne. 2011. "Testing the Future." The Alicia Patterson Foundation. http://aliciapatterson.org/stories/testing-future. Accessed on March 28, 2016.

Beera, Rajendra K. 2009. "The Story of the Cohen–Boyer Patents." *Current Science*. 96(6): 760–761.

Begley, Sharon. 1997. "Little Lamb, Who Made Thee?" *Newsweek*. March 10, 1997. Http://www.newsweek.com/id/95479. Accessed on March 28, 2016.

"Bioethics and Patent Law: The Case of the Oncomouse." World Intellectual Property Organization. http://www.wipo.int/wipo_magazine/en/2006/03/article_0006.html. Accessed on March 28, 2016.

Carroll, Dana. 2011. "Genome Engineering with Zinc-Finger Nucleases." *Genetics.* 188(4): 773–782. Available online at http://www.cpp.edu/~cwlamunyon/BIO421/Carroll11Genetics188_773.pdf. Accessed on April 3, 2016.

Colman, Alan. 1999. "Dolly, Polly and Other 'Ollys': Likely Impact of Cloning Technology on Biomedical Uses of Livestock." *Genetic Analysis: Biomolecular Engineering.* 15(3–5): 167–173.

Committee on Defining Science-Based Concerns Associated with Products of Animal Biotechnology. Committee on Agricultural Biotechnology, Health, and the Environment. National Research Council. 2002. *Animal Biotechnology: Science Based Concerns.* Washington, DC: National Academies Press.

Cox, David Benjamin Turitz, Randall Jeffrey Platt, and Feng Zhang. 2015. "Therapeutic Genome Editing: Prospects and Challenges." *Nature Medicine.* 21(2): 121–131.

Crick, F. H. C. 1958. "On Protein Synthesis." *Symposia of the Society for Experimental Biology: The Biological Replication of Macromolecules.* 12: 152–153.

"CRISPR-Cas9 and Genome Editing." 2016. Real Science and Other Adventures. https://realscienceandotheradventures .wordpress.com/2016/01/24/crispr-cas9-and-genome-editing/. Accessed on April 2, 2016.

Crommelin, Daan J. A., Robert D. Sindelar, and Bernd Meibohm, eds. *Pharmaceutical Biotechnology*, 3rd ed. New York: Informa.

"De-Extinction." 2016. National Geographic. http://www .nationalgeographic.com/deextinction/. Accessed on March 28, 2015.

Diamond v. Chakrabarty. 447 U.S. 303. 1980. Justia. U.S. Supreme Court. https://supreme.justia.com/cases/federal/us/447/303/case.html. Accessed on March 28, 2016.

Fernandez-Cornejo, Jorge, et al. 2014. "Genetically Engineered Crops in the United States." Economic Research Service. Economic Research Report Number 162. http://www.ers.usda.gov/media/1282246/err162.pdf. Accessed on April 1, 2016.

Fiala, Jennifer. 2005. "Researchers Create Mastitis-Resistant Cows." DVM360. http://veterinarynews.dvm360.com/researchers-create-mastitis-resistant-cows. Accessed on March 28, 2016.

Fraley, R.T., S.B. Rogers, and R.B. Horsch. 1983. "Use of a Chimeric Gene to Confer Antibiotic Resistance to Plant Cells." *Advances in Gene Technology: Molecular Genetics of Plants and Animals. Miami Winter Symposia.* 20: 211–221.

Framond, A.J., et al. 1983. "Mini-ti Plasmid and a Chimeric Gene Construct: New Approaches to Plant Gene Vector Construction." *Advances in Gene Technology: Molecular Genetics of Plants and Animals. Miami Winter Symposia.* 20: 159–170.

Frederickson, Jodi K. 1997. "He's All Heart . . . and a Little Pig Too: A Look at the FDA Draft Xenotransplant Guideline." *Food and Drug Law Journal.* 52(4): 429–451.

Gaj, Thomas, Charles A. Gersbach, and Carlos F. Barbas, III. 2013. "ZFN, TALEN and CRISPR/Cas-Based Methods for Genome Engineering." *Trends in Biotechnology.* 31(7): 397–405.

Gaxiola, Robert A., et al. 2001. "Drought- and Salt-Tolerant Plants Result from Overexpression of the AVP1 H^+-pump." *Proceedings of the National Academy of Sciences of the United States of America.* 98(20): 11444–11449.

"Genome Editing with CRISPR-Cas9." 2014. McGovern Institute for Brain Research at MIT. https://www.youtube.com/watch?v=2pp17E4E-O8. Accessed on April 2, 2016.

Hall, Harriet. 2009. "Recombinant Human Antithrombin— Milking Nanny Goats for Big Bucks." Science-Based

Medicine. https://www.sciencebasedmedicine.org/recombinant-human-antithrombin-milking-nanny-goats-for-big-bucks/. Accessed on March 28, 2016.

"History of Plant Breeding." 2015. My Agriculture Information Bank. http://agriinfo.in/default.aspx?page=topic&superid=3&topicid=2086. Accessed on March 27, 2016.

Holden, Simon. 2014. "Man Who Murdered Two Girls Aged 15 Was World's First DNA Conviction." *The Hinckley Times*. http://www.hinckleytimes.net/news/local-news/man-who-murdered-two-girls-7798455. Accessed on April 2, 2016.

"How the World's Governments Have Regulated Human Genome Editing." 2016. Bio-IT World. http://www.bio-itworld.com/2016/1/25/how-worlds-governments-have-regulated-human-genome-editing.html. Accessed on April 2, 2016.

Huffman, Wallace E. 2010. "Consumer Acceptance of Genetically Modified Foods: Traits, Labels and Diverse Information." Working Paper No. 10029. Iowa State University. http://ageconsearch.umn.edu/bitstream/93168/2/p11835-2010-08-10.pdf. Accessed on March 28, 2016.

Isasi, R., E. Kleiderman, and B. M. Knoppers. 2016. "Editing Policy to Fit the Genome?" *Science*. 351(6271): 337–339.

J. E. M. Ag Supply, Inc., DBA Farm Advantage, Inc., et al. v. Pioneer Hi-bred International, Inc.. 534 U.S. 124. 2001. http://caselaw.findlaw.com/us-supreme-court/534/124.html. Accessed on March 28, 2016.

Jaenisch, Rudolf, and Beatrice Mintz. 1974. "Simian Virus 40 DNA Sequences in DNA of Healthy Adult Mice Derived from Preimplantation Blastocysts Injected with Viral DNA." *Proceedings of the National Academy of Sciences of the United States of America*. 71(4): 1250–1254.

Leader, Benjamin, Quentin J. Baca, and David E. Golan. 2008. "Protein Therapeutics: A Summary and Pharmacological Classification." *Nature Reviews Drug Discovery.* 7: 21–39.

Levy, Marlon F., et al. 2000. "Liver Allotransplantation after Extracorporeal Hepatic Support with Transgenic (hCD55/hCD59) Porcine Livers: Clinical Results and Lack of Pig-to-Human Transmission of the Porcine Endogenous Retrovirus." *Transplantation.* 69(2): 272–280.

Lievens, A., et al. 2015. "Genetically Modified Animals: Options and Issues for Traceability and Enforcement." *Trends in Food Science & Technology.* 44(2): 159–176.

Liu, X, et al. 2014. "Generation of Mastitis Resistance in Cows by Targeting Human Lysozyme Gene to ß-Casein Locus Using Zinc-Finger Nucleases." *Proceedings of the Royal Society B: Biological Sciences.* 281(1780): doi: 10.1098/rspb.2013.3368. http://rspb.royalsociety publishing.org/content/281/1780/20133368. Accessed on March 28, 2016.

Lu, Dan, et al. 2016. "Large-Scale Production of Functional Human Lysozyme from Marker-Free Transgenic Cloned Cows." *Scientific Reports* 6, Article number: 22947. doi:10.1038/srep22947. http://www.nature.com/articles/srep22947. Accessed on March 28, 2016.

"Macrohistory: World History." 2016. http://www.fsmitha .com/index.html. Accessed on March 27, 2016.

Mawer, Mawer. 2006. *Gregor Mendel: Planting the Seeds of Genetics.* Chicago: Field Museum of Natural History.

"Meganucleases." 2010. Cellectis. https://www.youtube.com/watch?v=LV450LPTRDM. Accessed on April 2, 2016.

Melo, Eduardo O., et al. 2007. "Animal Transgenesis: State of the Art and Applications." *Journal of Applied Genetics.* 48(1): 47–61.

"Method of the Year 2011: Gene-Editing Nucleases." 2011. Nature Video. https://www.youtube.com/watch?v=zDkU FzZoQAs. Accessed on April 2, 2016.

Murai, Norimoto, et al. 1983. "Phaseolin Gene from Bean Is Expressed after Transfer to Sunflower via Tumor-Inducing Plasmid Vectors." *Science*. 222(4623): 476–482.

Murphy, Denis J. 2007. *People, Plants and Genes: The Story of Crops and Humanity*. New York: Oxford University Press.

National Human Genome Research Institute. 2009. "Cloning." http://www.genome.gov/25020028#7. Accessed on March 28, 2016.

"Pharming the Future." 2016. University of Calgary. http://www.ucalgary.ca/pharmingthefuture/pmf. Accessed on March 28, 2016.

Quill, Elizabeth. 2015. "These Are the Extinct Animals We Can, and Should, Resurrect." Smithsonian.com. http://www.smithsonianmag.com/science-nature/these-are-extinct-animals-we-can-should-resurrect-18095 4955/. Accessed on March 28, 2016.

Schell, J., M., et al. 1983. "Ti Plasmids as Experimental Gene Vectors for Plants." *Advances in Gene Technology: Molecular Genetics of Plants and Animals. Miami Winter Symposia*. 20: 191–209.

Sherman, David M. 2002. *Tending Animals in the Global Village: A Guide to International Veterinary Medicine*. New York: Wiley-Blackwell.

Stone, Brad. 2004. Press Release 94–10. U.S. Food and Drug Administration. Science Blog. http://www3.scienceblog.com/community/older/archives/M/2/fda1002.htm. Accessed on March 28, 2016.

"USPTO Patent Full-Text and Image Database." 2016. http://patft.uspto.gov/netahtml/PTO/search-bool.html. Accessed on March 28, 2016.

Waltz, Emily. 2014. "Transgenic Drought Tolerant Crops in Commercial Development and on the Market." From "Beating the Heat." *Nature Biotechnology.* 32(7): 610–613. http://www.nature.com/nbt/journal/v32/n7/fig_tab/nbt.2948_T1.html. Accessed on April 1, 2016.

Watson, J. D., and F. H. C. Crick. 1953. "A Structure for Deoxyribose Nucleic Acid." *Nature.* 171(4356): 737–738.

Wirt, Stacey. 2013. "Editing Our DNA with Molecular Scissors." The Tech. http://genetics.thetech.org/editing-our-dna-molecular-scissors. Accessed on April 2, 2016.

2 Problems, Issues, and Solutions

Introduction

Kem Ralph knew that what he was doing was wrong. The Tennessee farmer had agreed to help a neighbor disguise the cotton seeds he intended to plant in 2000. The seeds had actually been saved from a 1999 crop grown with a genetically modified product sold by the Monsanto company. When purchasing the seed originally, the neighbor had agreed not to save seed from the first year's crop to plant the following year, a practice that has been a tradition in farms around the world for centuries. Instead, Ralph agreed to tell Monsanto that the neighbor's seed came from his own crops. He, the neighbor, and many of their friends were angry at Monsanto's policy of restricting the use of their seeds for more than one season. Ralph's plan was discovered, however, and Monsanto decided to sue him for conspiracy to commit mail fraud. He was found guilty, sentenced to eight months in prison, and ordered to pay Monsanto the full value of the seed in question, $165,469, and a penalty of $1.7 million (Robinson 2009.)

Over the past half century, DNA technologies have made possible a host of new opportunities in agriculture, medicine, industry, forensic science, research, and other fields. Scientists now have the ability to change the most fundamental units of

A representative of the Monsanto company answers a consumer's questions about genetically modified corn at a conference in Austin, Texas. (Erich Schlegel/AP Images for GMO Answers)

which life is made, the molecules that make an artichoke an artichoke, and not a firefly; the molecules that determine how much milk a cow will produce; the molecules that determine whether a child will be healthy or fated to live with a devastating disease.

But these new opportunities have also created new challenges for scientists, public officials, and the general public. Some people criticize scientists for "playing God" by manipulating the units of which living organisms are constructed. And in doing so, they raise any number of moral, ethical, philosophical, theological, political, environmental, and social problems. Should plants be genetically altered in order to increase their productivity, if doing so may present a problem for human health and the environment? Should transgenic animals be created for the production of new kinds of foods if both humans and the animals themselves may be at risk from these procedures? Should medical scientists attempt to cure genetic disorders when the cures themselves may be a greater risk than the disease they seek to cure? Should individuals be tested for genetic disorders when test results may be misused by the government or private corporations and individuals may not themselves benefit from the information gained? These are only a few of the issues with which society in the 21st century is faced, given the advances in DNA technology.

Forensic Science

All new forms of technology inevitably face challenges from specialists in the field, from legal experts, and from the general public. Will the new technology function as effectively as its proponents claim? Are there risks to the technology that are not yet known? Is the new technology really superior to previous technologies? Questions like these always arise with a new invention or a new technology.

And such was the case with the use of DNA technology in forensic science. When Alec Jeffreys first introduced the possibility of using restriction fragment length polymorphisms (RFLP) for solving crimes in the early 1980s, forensic scientists and law enforcement officials wanted to know if the technology was really (1) as reliable as Jeffreys claimed and (2) superior to previous technologies, such as fingerprint analysis. The first legal challenge to the use of DNA technology in forensic science was posed in a 1989 case in New York state in which one Jose Castro was charged with the murder of a neighbor, Vilma Ponce, and her two-year-old daughter. At the ensuing trial, the prosecution offered evidence obtained from DNA tests that connected Castro with the murder weapon. The defense raised questions about the validity and reliability of the evidence, based in part on the fact that it was such a new technology. The trial judge did not rule on the testing procedure itself, but did bar the evidence because of human errors in the conduct of the testing by Lifecodes laboratory, the company that processed the DNA samples (Aronson 2007, Chapter 4).

Similar challenges have occurred on a number of occasions since *People v. Castro*, but they have almost always been based on errors in use of the test, statistical analysis of data, or other related issues, and not on the scientific principles on which the test is based. In fact, by 1994, two scientists who had previously been at odds over the use of DNA fingerprinting in forensic sciences came to the conclusion that "[t]he DNA fingerprinting wars are over . . . the public needs to understand that the DNA fingerprinting controversy has been resolved" (Lander and Budowle 1994, 735). One author (Budowle) had long been a supporter of the use of DNA typing in forensic science and had helped establish the FBI's DNA typing program, while the other author (Lander) had long been a critic of the use of this technology for the unambiguous identification of suspects. Over the years, however, the success of the technology had convinced Lander that his concerns were baseless, and the two men agreed that DNA typing was now entirely satisfactory as a valid and reliable method of identifying individuals. This conclusion does not mean, however, that DNA testing in forensic science is no longer a matter of dispute. In fact, the way in which evidence obtained by this technology is utilized remains an issue of concern to a number of specialists in the field and to the general public.

DNA Databases

One of the major questions of disagreement is the use of DNA databases. A DNA database is a collection of DNA samples taken from individuals and crime scenes stored at some location. Computer programs are written to compare DNA samples collected at the scene of a crime or taken from individuals arrested at the scene of a crime with samples taken from known criminals. It can also be used to solve so-called *cold cases*, crimes that law enforcement officials have been unable to solve by other means.

The first DNA database in the United States was established by the Federal Bureau of Investigation (FBI) in 1990 as a pilot program serving 14 state and local law enforcement agencies. The software program developed for that undertaking is called the *Combined DNA Index System* (CODIS). The success of the pilot program led to the passage of the DNA Identification Act of 1994 (now Public Law 103–322) that mandated the creation of a National DNA Index System (NDIS), a collaborative effort among the FBI and state and local law enforcement agencies. As of mid-2016, more than 190 public law enforcement laboratories participated in the NDIS, and CODIS software was being used by more than 90 law enforcement laboratories in over 50 foreign countries (CODIS Brochure 2016). The FBI points to the success of the NDIS/CODIS program by noting that the current database contains (as of February 2016) 12,205,768 offender profiles and 684,519 forensic profiles (DNA prints found at crime scenes), and it has produced over 322,011 hits (matches) in more than 309,614 investigations (CODIS—NDIS Statistics 2016).

These numbers are impressive, but they should not mask the fact that some controversy remains about the use of DNA databases. As originally conceived, these databases were to contain only DNA samples taken from crime scenes and from individuals convicted of violent crimes, such as murder, rape, and aggravated assault. They were also to be available for the identification of unidentified human remains and in the search for missing persons. Few objections arose about this arrangement. But as databases began to prove their value in obtaining convictions and connecting criminals with specific crimes, some law enforcement officials and politicians suggested expanding the reach of databases. They recommended including DNA samples taken from individuals arrested for or accused of misdemeanors, as well as felonies, whether those individuals were actually convicted of a crime or not. In 2003, for example, President George W. Bush recommended all adults and juveniles who had been arrested for a crime be required to submit a DNA sample (Willing 2003). Many state officials were also enthusiastic about expanding DNA databases. In 2007, for example, New York governor Eliot Spitzer proposed expanding the state's DNA laws to include the collection of DNA samples from anyone convicted of any crime, including misdemeanors such as minor drug offenses, harassment, or unauthorized use of a credit card. He would also have included anyone on probation, under parole supervision, or registered as a sex offender (McGeehan 2007). And in New York City, former mayor Rudolph Giuliani said that he would have "no problem" with fingerprinting all newborn babies in the city (Lambert 1998, B4).

A number of arguments have been put forward for expanding DNA databases. In the first place, law enforcement officers seldom know in advance who is going to commit a crime. The larger a DNA database, the more likely the DNA of a suspect will find a match in such a database. Second, anyone who is innocent of a crime or who has no plans to commit a crime should have no reservations about submitting his or her DNA to a database. Third, a very large database can help to exonerate innocent individuals accused of a crime. All that needs to be done is to compare DNA from a crime scene with the innocent person's databased DNA. Fourth, DNA evidence is a powerful tool not just in identifying criminals, but also in convicting them. Fifth, a large DNA database would probably be a deterrent to crime. A person might be less likely to commit a crime if he or she knew that his or her DNA information was already available to law enforcement officers. Sixth, DNA evidence is often easier to obtain and use than other forms of evidence. It can be obtained, for example, from a single drop of blood or semen, even after it has dried for weeks or months. Seventh, taking of DNA samples can hardly be considered an affront to civil liberties when the information obtained might be useful in preventing and solving terrible crimes.

A number of arguments have been posed against the expansion of DNA databases beyond the minimum absolutely required. Although the test itself is generally regarded as infallible, humans still make mistakes in conducting and interpreting those tests. Innocent people may still be convicted based on faulty DNA evidence. Second, the problem with any database is that once information has been entered, it generally remains there forever. It may not be possible to provide all the controls necessary to keep an individual's DNA information out of the hands of employers, insurance companies, medical companies, and other agencies who have no right to the information. Law enforcement officials must also note that they are not able to keep up with the processing of DNA samples already available to them (the waiting list sometimes runs many months or years), so how will they be able to process even more samples from less dangerous individuals? Perhaps the most troubling of recent recommendations for expanding DNA databases is that most people who are arrested for or accused of a crime are never convicted of that crime. Yet, having to provide DNA samples may well label those individuals as "suspicious," if not actually criminal. Finally, what kind of country would the United

States become if every citizen were required to provide a DNA sample, labeling him or her as a potential criminal?

DNA Fingerprinting Worldwide

The experience of many nations around the world with regard to DNA fingerprinting has been the same as that in the United States. Law enforcement officers have come to hold DNA testing in high regard, and the technology is widely acknowledged as being the "gold standard" in identification tools. As has also been the case in the United States, however, nations are struggling with the issue of DNA databases, attempting to decide how best to build such databases and how to use them in their own countries and in cooperation with other nations around the world. In 2016, Interpol, the international law enforcement agency, announced its DNA Gateway system contained more than 158,000 DNA records contributed by 73 nations ("Forensics" 2016). Most other nations around the world are either planning to develop a DNA database or are in the early stages of doing so. (For a complete list of DNA databases worldwide, see International DNA Databases 2016.)

One of the fundamental problems with the joint DNA database arrangement in Europe has been, as it has been in the United States, a decision as to which individuals should be included. Standards have varied widely across nations and have included any recordable offense (England); any crime with a potential sentence of more than eight years (the Netherlands) or more than two years (Sweden); any serious crime upon conviction (Belgium, Norway); certain crimes specified in DNA enabling legislation (France, Germany); any suspected recordable offense (Austria, Slovenia, England); any arrest for a serious crime or sexual crime punishable by more than a year in prison (Finland, Germany), by more than 1.5 years in prison (Denmark), by more than five years in prison (Hungary), or in association with a history of serious offenses (Switzerland). Nations also differ significantly on the issue of permission to remove DNA samples from a database. Some nations permit removal of a convicted offender's DNA specimen after a period of 5 to 20 years (Belgium, Denmark, Germany, Hungary, the Netherlands, Switzerland, Sweden, Slovenia), while other nations retain those specimens forever (Austria, England, Finland, Norway). Some nations remove DNA specimens from a database if a suspect is not convicted of a crime or never charged with a crime (Austria, Denmark, Finland, France, Germany, Hungary, Finland, and Switzerland), while England retains those specimens forever ("Introduction and Summary of Findings" 2011).

Low Copy Number DNA

A controversial issue in the use of DNA typing for forensic purposes involves a relatively new procedure known as *low copy number DNA* (LCN DNA), also known as *low template DNA* (LT DNA), *high sensitivity DNA* (HS DNA) "touch" DNA, and trace DNA testing. LCN DNA is a method that uses very small amounts of a sample of blood or other material (less than 200 picograms, or about a millionth of a grain of salt) for DNA analysis. The procedure is similar to that for larger amounts of a sample, but adds additional steps to the amplification process (which normally involves 28 steps) to 31 or more steps. As a consequence, investigators are able to obtain a DNA analysis from an object that someone has handled or a drop of blood that has dried months after it was deposited on a surface (Murphy 2015, 76–82; Word 2010).

Low copy number DNA analytical techniques were first developed by the Forensic Science Service in the United Kingdom, where it has been in use since 1999. It has also been adopted for use in criminal investigations by other law enforcement agencies, perhaps most prominently by the New York City Office of Chief Medical Examiner (OCME), where it was first employed in 2006. The technology is obviously of considerable value because it requires only the smallest imaginable amount of blood or other material in order to obtain a valid DNA test.

Still, it has come under considerable scrutiny primarily because it simply *is* so sensitive. Critics have pointed out that a sample taken for LCN DNA testing can easily become contaminated. If a police officer, for example, accidentally touches a sample taken at a scene crime—or the officer even shakes hands with another officer who has touched the sample—that sample may be contaminated with the officer's (or officers') own DNA, thus invalidating any later LCN DNA Test.

At the present time, LCN DNA testing is accepted by most courts as evidence in criminal cases, but questions remain. For example, in early 2016, Judge Mark Dwyer of the Brooklyn Supreme Court excluded evidence obtained by LCN DNA analysis because he believed the technology was not yet generally accepted by the scientific community ("Judge Dwyer Issues Written Decision in Landmark DNA Case Won by Legal Aid's DNA Unit" 2016). Concern about the technology has also been expressed by some of the individuals who have to use it. In early 2016, for example, a long-term employee at OCME, Dr. Marina Stajic, resigned from her post as director of the office's Forensic Toxicology Laboratory, allegedly because of her concerns about the reliability and validity of LCN DNA tests conducted at OCME (Weiser and Goldstein 2016).

Genetically Modified Organisms

Many people, both professional scientists and members of the general public, are very enthusiastic about the potential benefits of transgenic plants and animals in augmenting the nation's and the world's food supply. Probably most important to advocates of genetically modified (GM) foods is the potential for feeding untold millions of people around the world who will otherwise go hungry, become ill, or die from lack of food. Just as important is the fact, they say, that GM foods look and taste at least as good as conventional foods, have the same or better nutritional value, can be stored for longer periods of time, and are easier to ship. Farmers are often enthusiastic about

GM crops, because they produce greater yield on the same or smaller amounts of land and require less pesticides and other agrochemicals. The biggest problem with GM foods, proponents say, may be that people simply do not know enough about them and much of what they do know may be incorrect. Critics characterize GM foods as "frankenfoods," but they are actually very similar to conventional foods. Finally, advocates argue that the host of issues raised by opponents of GM foods—that they pose a hazard to human health and the environment—are not based on reliable and valid scientific evidence.

For those who oppose GM foods, however, the last point is often the most important; they believe that such foods do threaten human health and the natural environment, and they cite a number of specific instances as the basis for this concern. One such concern involves the fact that some individuals are allergic to certain foods, such as nuts, cow's milk, eggs, wheat, and soybeans. These individuals may experience reactions ranging from debilitating to fatal. A possibility exists that a genetically engineered food may contain a gene that might produce an allergic reaction in an unsuspecting consumer. In the 1990s, for example, Pioneer Hi-Bred International, Inc. experimented with the introduction of a gene from the Brazil nut plant (*Bertholletia excelsa*) into soybeans as a way of increasing the nutritional value of the soybeans. As part of its research, Pioneer commissioned a study of possible health effects of this engineered product. The study found that a small number of people who ate the engineered soybean experienced an allergic reaction to the product. Although the soybean was intended for use only as an animal feed, Pioneer decided not to proceed with the product (Leary 1996). Although this story would appear to have a happy ending, critics point to other instances in which GM foods have been implicated with allergic reactions and to the possibilities that such problems may occur in the future with inadequately tested GM products (for a comprehensive review of this argument, see "Genetically Engineered Foods May Cause Rising Food Allergies" [in two parts] 2007).

Another concern involves the development in humans of resistance to certain antibiotics. Researchers commonly add genes for resistance to an antibiotic (such as kanamycin) to an engineered plant or animal because that makes it easy to determine if the genetic transfer has been successful. But what happens if a food intended for consumption by humans carries a gene for antibiotic resistance? If that person later needs to take that antibiotic, the medication may not be effective for the person. Perhaps more important, the gene might then be transferred to the person's offspring, after which it could spread through the general population (Bodnar 2010; for an extended discussion of possible health risks to humans of GM foods, see "The Health Risks of GM Foods: Summary and Debate" 2016).

Critics of GM foods also point to a number of ways in which such crops can damage the environment. One concern is that genes used in the development of GM crops might escape into the surrounding environment and become integrated into related species, including weeds. In such cases, the weeds would themselves develop the same resistance to herbicides for which the GM crops were developed, and increasingly larger amounts of herbicides would be required to keep them under control. Thus far, there have been relatively few studies to suggest that such a phenomenon, commonly known as *horizontal gene transfer* (HGT), actually occurs in the field. One such report surfaced in 2002 when a British group reported that weeds growing adjacent to a field of genetically engineered oilseed rape crops in Canada had incorporated herbicide-resistant genes from the cash crop. The scientific advisor for the group, English Nature, warned that the ultimate effect of this case of HGT would be that farmers would have to begin using stronger, more dangerous pesticides to keep the weeds under control ("Rogue GM Plant Warning" 2002).

The troubling fact about this discovery is that evidence is now accumulating that HGT is not an uncommon event in nature, as was once thought. An initial hint at the scope of this problem appeared in a paper by a group of biologists at Indiana

University led by Jeffrey Palmer. That group found, "It appears horizontal gene transfer occurs for just about any gene in the plant mitochondrial genome" (Bergthorsson 2003; "Plant Genes Imported from Unrelated Species More Often than Previously Thought, IU Biologists Find" 2003). Five years later, the same researchers reinforced their original findings with another paper reporting on the very widespread occurrence of HGT in nature, not an especially propitious prospect for the wider use of GM crops (Keeling and Palmer 2008).

Critics of GM crops also worry about the effects of those crops on unrelated species in the environment, often referred to as untargeted species, because they are not the intended object of the DNA technology being used. One of the best-known studies in this field was conducted by a research team led by Cornell entomologist John Losey. That team removed pollen from genetically engineered corn and spread it on milkweed leaves that are the primary food of monarch butterfly larvae. It found that the larvae who fed on these leaves grew more slowly, ate less, and had a higher rate of mortality than did a control group of monarch larvae. The team concluded that the engineered corn posed a potential threat to the survival of monarch butterflies in areas where the crop was being planted (Friedlander 1999).

The next chapter in this story illustrates some of the problems in dealing with issues surrounding the use of transgenic organisms in agriculture. Six independent research teams followed up on the Losey study, all of them reporting their findings in the October 9, 2001, issue of the *Proceedings of the National Academy of Sciences of the United States of America*. The conclusion drawn from these studies was that the results obtained by Losey's team were not replicated in the field. Instead, the six teams agreed that, for a variety of reasons, the threat posed by the GM corn to monarchs in the field was negligible ("Bt Corn Pollen and Monarch Butterflies" 2001). The problem is that the follow-up studies received considerably less attention than did the original 1999 studies, so many people continue to believe the simplified version originally presented and perceive that GM crops are likely to pose a threat to other organisms in the environment (Food and Agriculture Organization. United Nations 2004, 71). That conclusion is certainly possible, but existing scientific studies suggest that the problem is probably less severe and more complex than it appears at first glance.

The availability of GM crops has transformed the face of agriculture around the world. As of 2013, 93 percent of all the soybeans grown in the United States, 82 percent of all cotton, and 85 percent of all corn came from genetically modified plants (Fernandez-Cornejo 2014, Table 3, page 9). These numbers mean that the vast majority of at least some crops being grown come from seeds that can be purchased from one source only: the company that manufactures those seeds. A farmer can not just go into the local feed-and-seed store and choose from half a dozen seed providers. The seed provider is likely to be Monsanto, which is now the world's largest supplier of genetically modified agricultural products, with more than 7,000 patents for biotechnological products, as of 2016 ("Results of Search in US Patent Collection db for: AN/Monsanto" 2016). One concern, then, is that the world's agriculture system is beginning to fall into the hands of a few very large transnational companies that will determine the character of crops that will be grown everywhere in the world.

A related problem is the degree to which these companies will ensure their control over crops. Most corporations that produce engineered seed require purchasers to sign an agreement that they will not save seeds from one year to be used in the following year (as was the case with Kem Ralph at the beginning of this chapter). These corporations argue that they spend hundreds of millions of dollars in the invention and development of engineered seeds, and it is only fair that they get a reasonable return on their investment. As a result, some companies now have "seed police" that go from place to place looking for violations of agreements that farmers have made with the company. When violations are discovered, as with Ralph, the individuals are likely to be prosecuted and, if convicted, fined, and perhaps imprisoned (Barlett and Steele 2008). Even seemingly innocent individuals may be at risk from the "seed police." In some cases, farmers have been prosecuted because seed from an adjacent plot of land has blown onto their own land, germinated, and grown to maturity. In such a case, those farmers have broken the law because they are growing a patented product (that blew in from the neighbor's field) for which they did not pay or sign an agreement with the manufacturer (Caruso 2004).

Fewer issues about the use of genetically modified animals have arisen, primarily because only one such animal has been approved for use as a source of food anywhere in the world, AquAdvantage Salmon, discussed later in this chapter. Still, some questions have arisen about other aspects of the genetic engineering of animals. One of the best known issues has surrounded the use of recombinant bovine somatotropin (rbST or rBST or rBGH, for recombinant bovine growth hormone) in dairy cows. Recombinant bovine somatotropin is a hormone implicated in the production of milk in cows. In November 1993, the FDA approved a petition from Monsanto for the use of a genetically engineered form of rBST in dairy cows. The purpose of rBST was to increase milk production in cows. The FDA found that rBST was safe and effective for dairy cows and safe for human consumption. The FDA also found that the use of rBST had no environmental effects and that there are no significant differences between milk from cows that have been treated with rBST and those that have not been treated (U.S. Department of Health and Human Services 1994, 6279).

In spite of this apparent robust endorsement of rBST use in dairy cows, a storm of protest has developed against the Monsanto product. One of the major complaints has been that milk from rBST-treated cows may have a higher-than-normal concentration of insulin-like growth factor-1 (IGF-1), a compound known to be implicated in the development of cancer in humans. Some critics argue that IGF-1 in rBST milk may be 40 times as high as that in untreated milk, vastly increasing the risk of cancer in humans (Kirk 2008). Thus far, studies appear not to have supported this view ("Report on the Food and Drug Administration's Review of the Safety of Recombinant Bovine Somatotropin" 2014). Nonetheless, a number of major retailers, including Starbucks, Kroger, Walmart, and Safeway, have decided not to sell milk from rBST cows.

The First Genetically Modified Animal: AquAdvantage Salmon

Research on the first genetically engineered fish to be approved by the FDA began in 1989 at the AquaBounty Technologies, in Waltham, Massachusetts. The fish was produced by inserting a growth regulating hormone gene taken from the Chinook salmon into Atlantic salmon, resulting in a time-to-maturity half that for the engineered fish as for its non-engineered cousins. All of the experimental and developmental work on the fish was conducted in an isolate pond kept under a high level of protection in Panama.

AquaBounty applied for FDA approval in 1995, and then waited for nearly 18 years before the agency finally acted on that request and granted approval for sale of the fish in the United States. The FDA's decision was based on the fact that the engineered salmon was not fundamentally different from its non-engineered cousins in any way, and it posed not risk to the environment. The FDA's decision allowed AquaBounty to begin market the AquAdvantage Salmon, although it by no means brought an end to the controversy as to whether it or other types of engineered plants and animals should be allowed into the public marketplace (AquAdvantage Salmon 2016; Goubau 2011; Ledford 2013).

Public Opinion about GM Foods

Researchers have been asking for nearly two decades what the general public thinks, in general, about the consumption of genetically modified foods. Over that period of time, two trends are noticeable. First, less than half of respondents to such surveys believe that GM foods are safe to consume. In the most recent poll by the Pew Research Center for the Internet, Science, and Tech, 37 percent of respondents said they thought GM foods *are* safe to eat, while 57 percent said that they were *not* safe to eat (Funk and Rainie 2015). Those numbers are roughly comparable to data collected by Pew researchers beginning in 2001 ("Search Results for 'Genetically Modified Foods'" 2016). Interestingly enough, members of the general public have a very different view about GM foods than do scientists themselves. In the 2014 Pew study, researchers compared the opinion of scientists about GM foods to those of the general public and found a startling difference: 88 percent of scientists believed that GM foods were safe to eat compared to 11 percent who said they were not. This difference between public and scientific views on the topic was the greatest difference for any socioscientific issue, such as climate change or abortion ("Major Gaps between the Public, Scientists on Key Issues" 2015). Perhaps even more interesting is the finding that a large majority of the general public believes that scientists still do not have a clear understanding of the health effects of GM crops, with 67 percent of Pew respondents expressing this view compared to only 28 percent who think that scientists are well informed on the subject (Funk and Rainie 2015).

The other point of interest is that the general public appears to be relatively uninformed about the scientific and technical issues involved in the development, production, sale, and consumption of GM foods. In some surveys, anywhere from 54 to 65 percent of Americans report that they have heard, read, or seen nothing about them ("Search Results for 'Genetically Modified Foods'" 2016). Yet, increased knowledge about GM foods appears to be fairly strongly related with attitudes toward their consumption. In the 2014 Pew survey, for example, people with knowledge about science in general tend to be evenly split between favoring and opposing GM foods, while those with little or no knowledge of science tend to oppose

GM foods by a margin of 66 percent to 22 percent (Funk and Rainie 2015).

Regulation of GM Crops and Food

The policies under which genetically modified foods are regulated differ from country to country. In the United States, the general principle that drives regulation is that GM foods have to meet the same standards as conventional foods. A detailed and complex system of regulation has developed to monitor the experimental testing of engineered seeds and the testing of genetically modified foods. But the standards used to approve GM foods are essentially the same as those used for any other new food produced in the United States ("Guide to U.S. Regulation of Genetically Modified Food and Agricultural Biotechnology Products" 2001; "How FDA Regulates Food from Genetically Engineered Plants" 2015).

Labeling of GM Foods

Critics of genetically modified crops and foods argue that the policy by which GM foods are regulated in the United States is too lenient and that more restrictive laws, like those adopted by the European Union, are more appropriate. Having failed thus far to accomplish this goal, U.S. critics have worked instead on legislation requiring that GM foods be labeled. They point out that people have the right to know what they are eating and, in particular, to know if the foods they buy contain genetically modified products. In 2006, Representative Dennis Kucinich (D-OH) introduced HR 5269, "The Genetically Engineered Food Right to Know Act of 2006," requiring the labeling of genetically modified foods. The bill was introduced and re-introduced in a variety of forms on seven occasions, but was never considered by the full House (H.R. 5269 (109th): Genetically Engineered Food Right to Know Act 2016).

Some observers have pointed out problems with the labeling of GM foods. They note that regulations already require food manufacturers to note on labels if a food differs in composition from a conventional food of the same type. But GM foods do not differ in composition from conventional foods, although they do differ in method of production. In addition, deciding which foods would have to be labeled might be an overwhelming problem. Given the complex composition of many foods today in the United States, it is likely that at least one component of a great many foods has been genetically modified in some way or another. After all, more than 80 percent of the corn grown in the United States has been genetically modified in one way or another. Finally, the cost of implementing a labeling program would be far greater than simply printing a new label. It would involve maintaining an extensive and complex system to decide what foods qualify for labeling and what goes on the label. Some authorities have estimated that a labeling program would add on average an additional 10 percent to the cost of labeled foods (McHughen 2008).

The debate over the labeling of genetically modified foods having gone essentially nowhere on the federal level, proponents of the practice have redirected their efforts to the state and local levels to achieve their objectives. Each year recently, a number of bills have been introduced into state legislatures (and, as often, at the local level) to require labeling of GM foods. In 2015, for example, 101 bills were introduced in 29 states on the topic of GM foods, of which 15 were passed. These 15 bills were only remotely concerned with labeling and, as of 2016, only one state, Vermont, has passed a bill requiring the labeling of GM foods ("State Legislation Addressing Genetically Modified Organisms" 2015).

At the same time, opponents of labeling have turned their attention to the U.S. Congress, hoping that they can obtain legislation that will be more favorable to food companies and prevent adoption of further laws at the state level, which have the possibility of being stricter than any federal legislation. An example of such legislation is H.R. 1599, introduced by Representative Mike Pompeo (R-KS) in the 114th

(2015–2016) Congress. The bill would require the FDA to have food labels note any difference between genetically modified and traditional foods if and when such differences were found to exist (which, so far, they have not). Pompeo's bill passed the House, but, as of mid-2016, has not been considered by the Senate (Congressional Train Rolling to Pre-empt States on GMO Labels 2016).

Public opinion polls have routinely shown that the general public is generally supportive of GM food labeling laws. A review of the most recent of those polls, for example, found that anywhere from 89 to 96 percent of respondents indicated that they favored the labeling of foods that contain genetically modified components. Polls going back to the early 1990s show a similar trend, although the favorable majority has not always been as significant as it is today (US Polls on GE Food Labeling 2016).

A number of food companies have become acutely aware of the pressures to label any GM foods they may produce. As with the U.S. Congress, some of those companies have decided that it might be better for them to create their own standards for GM food labeling and initiate such labeling before they are required to do so by a state or the federal government. In 2016, for example, the General Mills company announced that it would begin to label foods that contain genetically modified components, following the general instructions provided in the Vermont law. The first mock-ups of the company's label proved to contain only "relatively minor" changes, according to one observer, but did follow the guidance provided by the Vermont legislation (Blackmore 2016).

GM Foods in the European Union

Concerns about the cultivation, harvesting, transportation, and sale of genetically modified crops and foods dates to the early 1980s in the European Union (EU). Resolving these issues has been a particularly difficult challenge for the EU because of widely varying opinions about the safety and value of such crops and foods among the 28 member states. The first two documents produced on the issue were Directive 2001/18/EC, dealing with the deliberate release of GMOs into the environment, and regulation (EC) 1829/2003, on genetically modified food and feed in general. The two directives were largely unsuccessful as an effort to standardize and unify the approval process for genetically modified foods. While some countries attempted to follow the guidelines closely, other nations tended to ignore recommendations and develop their own systems for regulation and approval (GMO Legislation 2016).

Between 1994 and 1998, the European Union approved nine genetically modified crops, primarily varieties of corn, soybeans, and oilseed crops ("USTR Details Urgency of Ending EU Ban on Biotech Food" 2003). But individual nations had begun to express their own concerns about the future of GM foods. In 1997, for example, Austria banned a genetically modified variety of corn that had already been approved by the European Union. When the Union took no action on Austria's decision, other countries followed Austria's lead. In 1997, Luxembourg also banned an EU-approved corn, followed in 1998 by a French ban on two varieties of rapeseed and a Greek ban on GM rapeseed, in 1999 by an Italian ban on four varieties of corn, and in 2000 by a German ban on GM corn. In all cases, the banned products had already been approved by the EU itself ("USTR Details Urgency of Ending EU Ban on Biotech Food" 2003).

By June 1998, environmental ministers of the EU had adopted an informal, de facto ban on all GM foods. That ban applied both to the planting of genetically modified crops within the European states and to the import of GM foods to Europe. Those nations and individuals who support this ban have argued their case on the basis of the precautionary principle, which says that governmental agencies may be justified in taking certain regulatory actions even when some scientific uncertainty remains about the possible risks and consequences of a practice. In other words, "better safe than sorry" if the consequences of a practice are not well known and generally agreed upon. Over the ensuing decade, a number of efforts were made by the EU itself to invalidate the ban on GM foods in some member countries and to allow both planting and importation of such products, but all such efforts have thus far failed (Harrison 2009).

The inability of the European Union to remove bans on GM foods on the continent became an increasing source of irritation to the U.S. government, which argued that this inaction was a violation of free-trade agreements between the United States and the European Union. Those agreements, U.S. representatives claimed, prohibited bans by any nation on the free flow of products among signatories to the agreements. In early 2003, the administration of President George W. Bush filed a formal complaint about EU practices with the World Trade Organization (WTO). By the time the Bush administration took this action, however, the EU had already been forced into rethinking its position on GM foods. In March 2000, the European Court of Justice, the EU's highest judicial body, ruled that France had improperly banned three GM foods that had previously been approved by the EU ("Judgment of the Court 21–03–2000" 2000). Gradually, member states began to adopt the position that GM foods could be grown and imported provided they met certain labeling and traceability standards. These standards were ultimately expressed in directive 2001/18 of the European Community, whose two main provisions are that (1) foods containing more than 0.9 percent genetically modified organisms must be so labeled, and (2) all GM foods must contain a traceability tag, a piece of DNA that contains the "address" of the company by whom it was produced ("Directive 2001/18/ec of the European Parliament and of the Council of 12 March 2001" 2001). A traceability tag allows a governmental agency to track down the manufacturer of a product that is found to have some deleterious effect on human health or the environment (Papademetriou 2014).

In May 2004, the ban on GM foods in the European Union officially came to an end with approval by the European Commission of a GM corn made by the Swiss company Syngenta (Meller and Pollack 2004, C1). Two years later, the EU set standards for the acceptance of GM foods that included two requirements. First, each product submitted for approval had to be shown to be safe for humans, other animals, and the environment. Second, every product had to be available on a "freedom of choice" basis. That is, every farmer and consumer was to have complete freedom in deciding whether or not to use a GM crop or food. Manufacturers were required to provide data needed to allow people to make those decisions ("EU Alters GMO Assessments to Satisfy Resistant Member States" 2006).

Today, genetically modified foods and feed are readily available in the European Union, although there are still vigorous efforts under way to keep the continent "GMO-free." As something of an afterthought, the World Trade Organization finally ruled on the U.S. complaint of 2003 on May 16, 2006. The organization decided in favor of the United States and its co-complainants Argentina, Australia, Brazil, Canada, India, Mexico, and New Zealand, indicating that the European Union's long ban on GM foods and feed was an infringement of free-trade agreements among all parties. (For an overview of the current status of regulations on GM foods in the European Union as of 2016, see "Country Reports: GMOs in EU Member States" 2016.)

Xenotransplantation

The use of transgenic animals for xenotransplantation has raised another whole set of questions about the appropriate use of DNA technology in dealing with everyday medical problems. Xenotransplantation is the transfer of cells, tissues, or organs (but usually organs) between two species. In 2016, more than 121,000 people were waiting for a transplanted liver, kidney, pancreas, heart, lung, or intestine (United Network for Organ Sharing 2016). The supply of organs for transplantation from humans is far too small to meet the needs of all patients awaiting a transplant, and, as a result, hundreds of those patients die every year. The use of organs from other mammalian species, especially animals who have been genetically engineered to more closely match human DNA patterns, holds a very significant potential for saving thousands of lives every year.

Xenotransplantation raises some serious technical, ethical, and legal problems, however. Most fundamental are technical questions because, until they are solved, all other issues are essentially irrelevant. As noted in Chapter 1, efforts are now under way to engineer pigs and other animals by introducing genes that reduce the likelihood of rejection of a transplanted organ by a human host's body. This step alone has significantly improved the chance of success of a xenotransplant. Other technical problems remain, however. One of the most serious is the risk that a transplanted organ will carry with it a virus from the donor animal that then grows and multiplies within the human host body. Such an event would pose a threat not only to the host, but also to his or her progeny, especially since some viruses remain dormant for many years. When the FDA proposed draft guidelines for the use of xenotransplantation, this concern drew the largest number of comments from virologists, other members of the medical and scientific communities, individual citizens, representatives of the American Society of Transplant Physicians and the American College of Cardiology, and commercial sponsors of xenotransplantation clinical trials (U.S. Department of Health and Human Services. Food and Drug Administration. Center for Biologics Evaluation and Research 1999, 2–3).

Another common objection to the use of animals for xenotransplantation is based on the principle that non-human animals also have rights. As one critic of the procedure has said,

Primates used in xenotransplantation research will experience a large number of traumatic procedures, including major surgery, from which many will die; internal haemorrhages; isolation in small cages; repeated blood sampling; wound infections; nausea, vomiting and diarrhoea because of immunosuppressant drugs and kidney or heart failure. . . . This is a terrible waste of animal lives. (Bryan and Clare 2001)

The most common response to this concern is that humans already keep and slaughter animals for food, clothing, and any number of other purposes, so using them to save human lives is not really a new use of non-human species.

Pharming

For over two decades, pharmaceutical companies have been excited about the possibility of using plants and animals for the production of drugs. As described in Chapter 1, it is possible to insert the genes needed for the synthesis of almost any given protein into the DNA of a cow, goat, sheep, pig, tobacco plant, or other organism, such that that animal or plant becomes a "bioreactor," a living factory for the production of that protein. The desired protein can be harvested from the plant or animal in a number of ways, for example, by collecting an animal's urine, blood, or milk. The protein is then separated from the fluid by chemical means and purified. The final product is normally identical to the protein made by chemical means in the laboratory.

As of 2016, pharming with animals has been sufficiently developed to the point that a number of essential drugs are now available using this technology. They include Factors VII, VIII, and IX, for the treatment of hemophilia; fibrinogen, for the treatment of congenital fibrinogen deficiency; collage, for tissue repair, lactoferrin, for nutritional deficiencies, and antithrombin III, for the prevention of thrombotic events (Kayser and Warzecha 2012, Table 5.1, 75).

Commercially, animal pharming poses something of a problem because of the initial costs of developing a suitable bioreactor. Only about 1 percent of the eggs that received transformed DNA will, on average, result in a live birth, and only a fraction of those animals will actually express the transcribed gene ("Pharming for Farmaceuticals" 2016). Simply producing a single animal with the ability to produce the desired drug, then, can be a hugely expensive challenge. The upside of that story, however, is that once the animal is successfully produced and itself begins to reproduce, the bioreactor system is established. From that point on, other than maintaining the transgenic animals, the desired drug is produced at almost no cost to the pharmaceutical company. A recent review of the costs of producing proteins using animals compared to alternative methods (such as the use of prokaryotes) concluded that, when all costs of production are taken into account, animal pharming has a financial advantage over other available systems of production.

The use of transgenic plants for the production of desired products has been at least as successful as that with animals. The first such product made this way was human serum albumin, produced in 1990, from transgenic potato and tobacco plants (Sijmons et al. 1990). Since that time, a number of other useful proteins have been made commercially available using transgenic plants, including the enzyme glucocerebrosidase, for the treatment of Gaucher disease; the enzyme gastric lipase, for the treatment of cystic fibrosis phase; alpha interferon, for the treatment of hepatitis B and hepatitis C; human intrinsic factor, for the treatment of vitamin B12 deficiency; insulin, for the treatment of diabetes; and lactoferrin, for the treatment of dry eye syndrome and gastrointestinal disorders (Rehbinder et al. 2008, Table 2.1, 12).

One of the most intriguing fields of research in pharming involves the development of edible vaccines. Injected vaccines have for many decades been a key tool of public health workers in protecting children and adults from potentially serious diseases, such as diphtheria, hepatitis, polio, typhoid fever, ty-phus, and yellow fever. But injectable vaccines have a number of drawbacks. Many people avoid being vaccinated because of their fear of needles. Delivering needed vaccines to their target populations may be time-consuming, and the vaccines may lose potency while being stored. They may also be destroyed by exposure to high temperatures. And simply because of the way in which they are delivered—by injection—they do not necessarily provide the maximum protection possible.

Transgenic plants may also hold the promise for the manufacture of novel pharmaceuticals or of products not available by other means. One of the most frequently cited examples is edible vaccines. Some scientists are now hopeful that edible vaccines can be produced that overcome all or most of these obstacles. Genes coding for parts of human pathogens (but not the whole pathogens) are inserted into the DNA of edible plants, such as bananas, potatoes, or rice, and the fruit of those plants is then fed to humans. Inside the human body, the pathogen units stimulate an immune response that is characteristic of vaccines, providing the same or better protection than do injectable vaccines (Langridge 2000; Yu 2008).

Currently, no edible vaccines for use with humans have yet been approved by the FDA. However, a number of such vaccines have been developed and approved for use with domestic animals, primarily chickens (for Newcastle disease and infectious bronchitis), pigs (enterotoxigenic *E. coli* disease, foot and mouth disease, and transmissible gastroenteritis), and cattle (bovine herpesvirus and rinderpest virus) (Takeyama, Kiyono, and Yuki 2015; "Vaccine Development Using Recombinant DNA Technology" 2008).

Edible vaccines have their critics, as do most other products of transgenic plants and animals. The primary concern seems to be that pollen, seed, dust, or plant parts from a transgenic plant might escape from a field and get carried to nearby crops or private gardens. These plant materials could then be absorbed by soil microbes, eaten by farm animals, and/or eventually ingested by humans. The risk posed by this possibility depends to a great extent on the vaccine being produced in the transgenic plant. If, for example, the pharmaceutical being made were a blood thinner, it might pose a threat to the health of humans or animals by whom it was ingested. At this point, more information is needed about the possible combined threat posed by escape of the engineered vaccine and risk posed to human and animal health (Cummins 2014; "Eat Up Your Vaccines" 2000).

In the United States, the regulation of commercial products made by recombinant DNA technology is controlled by a set of guidelines developed by the Office of Science and Technology Policy in 1986. The basic principle guiding the development of these guidelines was that products approved for commercial use should be safe for use by the general public without the necessity of imposing burdensome rules and regulations on biotechnology companies. The specific document that evolved from these principles is very complex and is challenging for the average person to read and understand. It parcels out responsibility for regulating genetically modified products to three agencies, the Food and Drug Administration (FDA), the Environmental Protection Agency (EPA), and the Department of Agriculture's Food, Safety, and Inspection Service (FSIS) and Animal and Plant Health Inspection Service (APHIS).

For example, the regulation of human and animal drugs is the responsibility of the FDA, while foods and food additives are the joint responsibility of the FDA and FSIS. Regulation of pesticide microorganisms released to the environment is the joint responsibility of the EPA and APHIS ("Coordinated Framework for Regulation of Biotechnology" 1986). The Framework was revised and updated in 1992. The major feature of the 1992 revision was to emphasize that the primary focus of regulation of genetically modified materials was on (1) the characteristics of the product itself, (2) the environment into which it is to be introduced, and (3) the intended use of the product, rather than upon the process by which the product was made. Since the U.S. government at the time had already concluded that GM crops and foods were essentially indistinguishable from their non-GM alternatives, this rule established the same standards for GM products as for non-GM products (Belson 2000).

In 2015, President Barack Obama announced that he had ordered another review of the framework. Although, he assured the public, the framework had worked well in protecting the safety of engineered products for the general public, the framework itself was so complicated that it was difficult for the average person to understand what regulations provided for, and which agencies were engaged in what aspects of the regulatory process (Pollack 2015).

Genetic Testing

The term "genetic testing" actually refers to a number of different procedures, including diagnostic testing, predictive and presymptomatic testing, carrier testing, prenatal testing, preimplantation genetic diagnosis, and newborn screening. Diagnostic testing is used when a medical professional has reason to believe that a person has or is likely to develop some form of genetic illness. The testing is done to confirm that diagnosis, improving the ability of a person to choose treatments and other options to prolong his or her life or to make life more comfortable. An example is the test for a disease known as hemochromatosis, a genetic disorder in which a person's body accumulates iron over long periods of time. Excess iron in the body eventually results in a variety of physical problems that may include cardiac dysfunction, cirrhosis, diabetes, and liver cancer. If the problem is diagnosed soon enough, some simple procedures are available for reducing the risk of such problems. One approach is phlebotomy, or bloodletting, in which a certain amount of blood is removed from the body from time to time. Diagnostic tests are now available for the presence of the defective gene that causes hemochromatosis, making it possible for a medical worker to counsel a person about the best way to deal with this disorder. As with all diagnostic genetic tests, the test for hemochromatosis can be conducted at any time in a person's life or prior to birth.

Predictive and presymptomatic tests are used with individuals who (usually) have demonstrated no signs of a disease and/or who have reason to believe they may be at risk for a genetic disease. One of the classic conditions for which such tests are used is Huntington's disease (HD), once called Huntington's chorea. Huntington's disease is a genetic disorder that affects the nervous system, producing a loss of coordination, unsteady gait, and jerky body movements, accompanied by a decline in mental abilities and psychiatric and behavioral problems. The condition is incurable. Perhaps the most troubling aspect of HD is that it typically does not begin to manifest until relatively late in life, after age 40. Predictive and presymptomatic genetic testing makes it possible for a person with a family history of HD to know at any early stage of life whether he or she has the gene responsible for HD and, thus, is likely to develop the disorder. This information makes it possible for a person to plan in whatever way possible for living with and treating the condition.

Carrier testing is used to detect recessive genetic disorders. Some genetic disorders are said to be recessive because they are expressed only when a person inherits a defective gene from both parents. If the person inherits a defective gene from one parent and a normal gene from the other parent, the person does not develop the disease, but is only a carrier for the disease. Knowing that one has one defective copy of the gene and one normal copy of the gene has relatively little significance to an individual; he or she will remain healthy with this genetic background. But this information can be very important to a couple planning to have children. If both parents have one defective copy of the gene (are carriers for the disorder), there is a one-in-four chance that any children they have will receive two copies of the defective gene and will develop the genetic disease. This condition holds for a number of genetic disorders, including cystic fibrosis, Niemann-Pick disease, sickle-cell anemia, spinal muscular atrophy, and Tay-Sachs disease. With the information from carrier genetic testing, a couple can decide whether or not to have children and/or plan how they would care for a child with the designated disorder.

Prenatal testing is often recommended for pregnant women who fall into one of four categories: those who are over the age of 35 (because of increased risk for fetal chromosomal abnormalities); those who have a family history for some genetic disorder, such as cystic fibrosis or Duchenne muscular dystrophy; those whose ethnic or racial background might predetermine a person to a genetic disorder, such as Tay-Sachs disease or sickle-cell anemia; and those who are concerned about the possibility of some other type of genetic disease. Prenatal testing is usually conducted by amniocentesis or chorionic villus sampling (CVS). In either case, a small sample of fetal material is removed and examined for genetic or chromosomal abnormalities. The results of these tests help couples and their health providers to consider birth and care options for the unborn fetus and, later, the newborn child.

Preimplantation genetic diagnosis (PGD) is a relatively new type of genetic testing. Women who hope to become pregnant can have eggs that have been fertilized in vitro (outside the human body) examined for possible genetic or chromosomal defects before they are implanted. Before the availability of PGD, the only way to test for such abnormalities was to proceed with the implantation and then use amniocentesis or CVS at some later date to look for possible fetal defects.

"Newborn screening" is a term that refers to a testing program used with every newborn child, whether any reason exists for concern about possible genetic disorders or not. A screening program is used to identify certain somewhat common disorders that can and should be treated as soon as the child is born, phenylketonuria (PKU) being perhaps the best known example. Left untreated, PKU can be fatal. But it can be treated very easily by providing a newborn child with a specialized diet, and the child can then live a normal life as long as he or she continues on that diet. All states now require some kind of newborn screening panel in which tests for a minimum of 26 disorders are conducted ("Newborn Screening Tests" 2015).

Risks and Benefits

There is probably no question that genetic testing can have very significant benefits for parents and children. The detection of PKU is perhaps the best single example. If the condition is detected before a child is born, dietary adjustments can be made in her or his life that make a potential life-threatening condition into an easily controlled disorder. Another example is the genetic condition known as familial adenomatous polyposis (FAP). Individuals with this condition develop numerous benign polyps (growths) on their colon, often beginning as early as their teenage years. Over time, these polyps tend to become cancerous, resulting in a life-threatening condition. If an individual knows in advance that she or he carries the genes for FAP, steps can be detected to reduce or eliminate the severity of the disorder ("Familial Adenomatous Polyposis" 2013).

Still, a number of concerns exist about the use of genetic testing for pregnant women, newborn children, and adults. In the first place, the possibility of error always exists in genetic tests, as it does in any human endeavor. Chemicals may not react in the manner expected (and the manner they do 99 percent of the time); humans may make errors in performing a test, or they may make errors in reading and interpreting the test. Such errors can have profound results. A woman when told that her unborn child carries the gene for cystic fibrosis, for example, might decide not to continue the pregnancy. But what if that result were erroneous? The woman's decision would have been based on incorrect information and would have resulted in an act to which she would otherwise never agree.

A second problem concerns the usefulness of information obtained from a genetic test. The number of monogenic disorders (diseases caused by an error in a single gene) is relatively small, and even monogenic disorders can be complex and unpredictable. For example, more than 300 different mutations are possible in the gene that causes cystic fibrosis, and each mutation produces a disease of differing severity and characteristics. To discover that one is carrying an erroneous gene for cystic fibrosis provides, therefore, only minimal information for its possible treatment (Andrews et al. 1994, 62).

Arguably the most common issue related to genetic testing today involves testing for breast cancer. Genetic factors are implicated in a significant number of cases of breast cancer, many of which are related to mutations in two genes: BRCA1 and BRCA2. These genes are responsible for the production of proteins that suppress the growth of tumors in the breasts. If mutations occur in the genes, the body's natural system for fighting breast tumors is diminished. Genetic testing makes it possible for a woman to learn whether or not she has such mutations in the two BRCA genes. If she does, she has an elevated risk for breast cancer, to which she can respond in a variety of ways. One such response might be surgery in which one or both breasts are removed, greatly reducing the likelihood of cancer's developing at a later stage in life. The consequences of such surgeries are, of course, profound, and deciding how to deal with positive genetic test results can ultimately be one of the most difficult issues a woman has to face in her life ("BRCA1 and BRCA2: Cancer Risk and Genetic Testing" 2015).

How valuable is the information, then, that one is carrying one or more genes for some incurable disorder, such as some types of breast cancer, cystic fibrosis, or Huntington's disease? Granted, such information allows a person to make lifestyle choices, financial decisions, end-of-life choices, family planning, and other arrangements about one's future. But at what emotional cost? Imagine the turmoil experienced by a 20-year-old man who has just tested positive for Huntington's disease, which may not manifest for 20 years or more. The man knows that his disease, if and when it develops, is incurable, but he also now has the option to decide whether to have a family and how to plan for his future. Is he better or worse off for having had the genetic test? Each individual will answer that question differently.

Issues of Privacy and Confidentiality

Another common concern about genetic testing arises over questions of privacy and confidentiality. There is probably nothing more obviously a person's own private business than her or his genetic information. How many people are willing to share with outsiders basic data about their own genome? Genetic testing companies usually take extensive precautions about the genetic data they collect from people. Yet, those data can easily fall into the hands of outsiders, most commonly insurance agencies. In order to issue a health or life insurance policy to an individual, a company may want to know (understandably) everything possible about a person's health condition. Insurance companies are typically not eager to issue policies to people who have fatal illness or are at risk for serious health problems. If they know a person has had genetic tests, they will probably want to know the results of those tests, and not insure those with identified long-term serious health problems. Is this a legitimate use of information obtained from genetic tests?

One of the most recent issues related to genetic testing has been the use of testing kits designed for consumer use. Such kits are sold on the Internet or through other outlets directly to consumers. They usually have the consumer take a DNA sample (e.g., by swabbing the inner surface of one's cheek) and send that sample to the sponsoring company for analysis. These kits commonly carry clear and specific warnings to the effect that they are not designed for making specific diagnoses. In 2006, the U.S. General Accountability Office (GAO) conducted a study to assess four testing kits purchased on the

Internet. They submitted cheek swabs and complete medical histories from 14 fictitious individuals to the four companies from whom they had purchased the kits. (All 14 swabs were actually obtained from a nine-month-old female.) The GAO study found, first of all, that all four kits they ordered online did contain adequate warning statements, such as "[This is] not a genetic test for disease or predisposition to disease, nor does it determine a medical condition," and "Please note that this screening is not a test for inherited disorders" (U.S. Government Accountability Office 2006, 8). But the study also found that all four companies reported that the 14 fictitious individuals were at risk for developing one or more genetic disorders, including cancer, osteoporosis, type 2 diabetes, high blood pressure, and heart disease. The companies then recommended a variety of treatments for the prevention and/or treatment of these conditions. In most cases, those treatments involved the use of dietary supplements, which the companies offered at costs of up to 40 times the expense of the same supplements available at local stores. The GAO concluded from its study that "all the tests GAO purchased mislead consumers by making predictions that are medically unproven and so ambiguous that they do not provide meaningful information to consumers" (U.S. Government Accountability Office 2006, [I]; for more recent information on genetic testing kits, see "Regulation of Genetic Tests" 2015).

Regulation

Until recently, only relatively modest federal legislation existed dealing with genetic testing. Standards for clinical laboratories in general has long been under the purview of the Food and Drug Administration, but there has been no specific regulation of genetic testing on a national level. That situation changed in 2008 when the U.S. Congress passed and President George W. Bush signed the Genetic Information Nondiscrimination Act, which prevents employers and health insurance companies from using information obtained from genetic tests to discriminate against individuals. The bill received virtually unanimous support in the Congress, passing the Senate by a vote of 96 to 0 and the House by a vote of 420 to 3, with nine abstentions. Proponents of gene testing regulation were pleased with these results. Senator Ted Kennedy (D-MA), for example, called the legislation "the first major new civil rights bill of the new century" (Gruber 2009).

In spite of the high praise and widespread support for the bill, some criticism of the legislation remains. One concern is that the bill is not as inclusive as some observers would like. For example, it does not protect people from discrimination when applying for life insurance or for long-term care or disability insurance (Keim 2008). Employer organizations such as the National Association of Manufacturers, National Retail Federation, Society for Human Resource Management, and U.S. Chamber of Commerce argue that provisions of the bill may be too onerous for many companies, requiring them to offer insurance to many people who have serious health problems. They also are concerned about possible conflicts between the new federal law and the many different state laws dealing with genetic testing (Fletcher 2007).

Most states have also passed legislation protecting the privacy of individuals who have had genetic tests. In most cases, those laws require that an individual give specific and informed consent for the use of information obtained from genetic testing to any other person or agency. These laws usually also prohibit a third party from requesting or requiring that an individual have a genetic test. A number of states also have laws that define DNA samples and/or the information they contain as personal property, allowing them to be considered within that category for legal purposes. Four states—Delaware, Nevada, New Mexico, and Oregon—specifically limit information from genetic tests to individuals only ("Genetic Privacy

Laws" 2008). One of the simplest laws is that of Hawaii, which has three provisions: (1) that genetic information cannot be used in making any decision as to whether a person is eligible for health insurance or the type of insurance that can be offered; (2) an insurance company cannot request or require that an applicant for insurance have any kind of genetic test; and (3) information obtained from genetic testing cannot be given to any other person or organization without an individual's written approval (see also Katz and Schweitzer 2010).

Genetic Counseling

The availability of genetic testing has spawned the development of a new career field: genetic counseling. The purpose of genetic counseling is to provide professional assistance to individuals and couples who are contemplating having genetic tests for one or more disorders or who have already had such tests and wish to better understand what those tests mean and what their possible implications are. Professional genetic counselors typically have earned a master's degree in medical genetics and counseling and have passed a certification examination administered by the American Board of Genetic Counseling (American Board of Genetic Counseling 2016).

Some people choose to meet with genetic counselors before they become pregnant; some, after they have become pregnant, but before they have had any genetic tests performed; and some, after tests have been completed and results are available. In each case, a counselor can review the nature of any one or more genetic tests, the kind of information produced by the tests, the relative seriousness of conditions indicated by positive test results, the likelihood that a positive test result indicates that a child will actually develop a given condition, pre-birth options for parents of the fetus, post-birth treatments and maintenance procedures that may be available, and other relevant issues that parents may have to deal with if a child carries a gene or chromosome associated with some specific disorder ("Making Sense of Your Genes: A Guide to Genetic Counseling" 2008).

Gene Therapy

As with other forms of DNA technology, the primary issue relating to human gene therapy (HGT) with which practitioners and the general public have to deal is technical problems. The early successes of HGT experienced by W. French Anderson and other researchers in the late 1980s and early 1990s (see Chapter 1) inspired a great deal of hope among researchers and patients with potentially fatal or severely disabling disorders and their families. A few deaths and other disastrous results, however, clearly highlighted the technical problems involved in treating genetic disorders as readily and successfully as early pioneers may have hoped. A number of companies working in the field of HGT simply gave up on the technology, while others have struggled through difficult times because of diminished hopes for the technology. Today, however, a number of experts in the field are feeling especially optimistic about the possible future applications of gene therapy in dealing with a number of medical conditions (DeWeerdt 2014).

Should technical problems involved in the use of gene therapy be solved and the technology become more generally available to individuals with severe genetic disorders, it seems possible that considerable public support for the procedure will develop. It seems somewhat difficult to imagine, absent technical risks for patients, how an individual would refuse a treatment that would save his or her life or vastly diminish the medical problems required simply to stay alive. Even then, however, some ethical questions remain about the use of gene therapy for the treatment of genetic disorders. Perhaps the most frequently mentioned of these issues is the so-called "slippery slope" argument.

The slippery slope argument says, in general, that allowing some action to occur in the present, even if that action seems harmless and benign, may set into action a series of other events with far less benign consequences. That is, the first step at the top of a slippery slope may not pose any problems whatsoever, but once that step is taken, there may not be any way to avoid a whole host of less pleasant circumstances, such as sliding all the way to the bottom of the slope.

In dealing with gene therapy, the slippery slope issue usually focuses on the distinction between therapeutic and enhancement (cosmetic) treatments. A therapeutic treatment is one designed to cure a disease, such as cystic fibrosis or sickle-cell anemia. An enhancement treatment is one intended to improve one's overall physical or mental condition, that is, making a person better looking or more intelligent (or, conceivably, more personable, more aggressive, more intuitive, more friendly, and the list goes on). It might appear that the two kinds of treatments are very different from each other. But such is not necessarily the case.

How would a person with mild allergies be classified? Those allergies can probably be controlled by avoiding the allergens responsible for one's condition. But if a genetic component of the allergic reaction could be found, would not gene therapy be a better long-term solution for the person's problem? Or what about a child who fails to grow normally (whatever "normal" might mean in this case)? If a defective gene responsible for this problem could be found, should not the person be offered the option of having gene therapy to "cure" the problem? In any case, many people clearly feel that certain physical characteristics can be a handicap in life and are willing to pay to have those characteristics altered.

Trends in Gene Therapy

In 2015, nearly 16 million cosmetic surgical procedures were conducted in the United States, ranging in price from relatively simple laser hair removal (at an average cost of $254) to lower body lifts (at an average cost of $7,958). The most common nonsurgical cosmetic procedures conducted were botox injection (4,267,038 cases), hyaluronic acid injection (2,148,326 cases), laser hair removal (1,136,834 cases), and chemical peel (603,305 cases). The most common surgical procedures were liposuction (396,048 cases), breast augmentation (305,856 cases), and tummy tucks (180,717 cases) ("Cosmetic Surgery National Data Bank Statistics" 2015).

With this level of interest in cosmetic surgery, some authorities are concerned that gene therapy will eventually become used for dealing with almost any "defect" that a person might imagine. David King, a former molecular biologist and director of the watchdog group Human Genetics Alert, has observed that:

> The fact is that science is not merely changing the barrier between disease and "normality": it is changing our whole conception of what it is to be human. There is an increasing reluctance to accept any predetermined limits of our biology and a tendency to regard human body as being perfectible. . . . When gene therapy becomes familiar, we will start to see our genes as less finalised. People are increasingly coming to feel that they are no longer stuck with the body they were born with. The definition of the human condition has now become a function of technology. . . . (King 1997; see also Harris and Chan 2008)

King has also expressed concern about another potential use of gene therapy. In general, gene therapy can be conducted along two lines: using somatic cells as the recipients of engineered genes, or using germ line cells. Somatic cells are any cells other than those that make up the reproductive system. Any change made in a somatic cell affects the organism into which it is introduced, but not its descendants. By contrast, genetic information contained in germ line cells (sperm and eggs, for example) is passed on from generation to generation, so that any modification made in such cells becomes part of the genome of succeeding generations. The only medical reason for modifying germ line cells is to stop the transmission of a genetic disorder from one generation to the next and, perhaps, to eliminate the disorder from the human race entirely. Thus far, for a variety of reasons, there has been virtually no research on the genetic modification of germ line cells. However, the medical potential of treating genetic disorders by germ line modification has been discussed seriously by researchers for at least a decade.

Some of those researchers believe that the use of germ line, rather than somatic, cells is "an ideal form of gene therapy" (Fox 1998). That may well be the case in the pursuit of methods for ending genetic disorders, but is it as acceptable for use in enhancement therapy? Finally, the most powerful argument against gene therapy among some people is that it represents an effort on behalf of scientists to modify the very nature of human beings. For these individuals, gene therapy is nothing other than "playing God," taking over responsibilities that are not allowed to humans. This argument quickly becomes very complex, as it is perfectly obvious that humans act as if they were "playing God" all the time. When a doctor acts to cure a disease, for example, he or she appears to be interrupting a fate that has been decreed by a higher power, if one does believe in such a power. In the case of genetic therapy, the problem is made difficult because some ethicists argue that somatic genetic therapy is acceptable because it is a curative act, while germ line genetic therapy is forbidden because it alters the very nature of human beings ("Genetic Engineering" 2009).

The ethical debate over gene therapy appears to have reached its peak in the 1990s when breakthroughs in the technology were anticipated to occur at any moment. In fact, progress in the field has been much slower than many experts believed at the end of the last century, and there appears to be less concern about moral and ethical issues today and more interest in a variety of technical problems. Still, an optimistic view is that the potential for human gene therapy will one day be realized, and, at that point, the moral and ethical dilemmas outlined here will once again come to the forefront (Szebik and Glass 2001, 32–38).

Regulations

The U.S. Food and Drug Administration is the lead federal agency in the regulation of human gene therapy research and development activities. In October 1993, it issued a general directive defining in detail its statutory authority for regulating various aspects of the development, testing, and use of new gene therapy technology (U.S. Department of Health and Human Services. Food and Drug Administration 1993). Since that time, the agency has continued to issue directives dealing with specialized aspects of research in the area of human gene therapy. Some examples of the topics covered in these directives are Draft Guidance for Industry: Potency Tests for Cellular and Gene Therapy Products, Guidance for Industry: Gene Therapy Clinical Trials—Observing Subjects for Delayed Adverse Events, Guidance for Industry: Guidance for Human Somatic Cell Therapy and Gene Therapy, and Guidance for Industry: Eligibility Determination for Donors of Human Cells, Tissues, and Cellular and Tissue-Based Products ("Cellular and Gene Therapy Products" 2015). Of special interest to the general reader are occasional articles on the topic for the general public. (See, for example, Thompson 2000 and Witten 2007; for all current FDA articles on gene therapy issues, see "Gene Therapy" 2016.)

Cloning

The term *cloning* refers to any process by which an exact copy of a gene, cell, organism, or other entity is made. Cloning is a very common natural process. When a bacterial cell reproduces asexually, for example, it splits into two genetically identical daughter cells. Those daughter cells then split again, into four new, but genetically identical, cells. The four split into eight, and so on. The same process occurs at the early stages of growth in all plants and animals. A fertilized human egg, for example, divides into two daughter cells, each of which is identical to the original cell, a process that is repeated many times over. This process, by which a group of genetically identical cells are produced from a single parent cell, is called *cell cloning*.

Cell cloning that occurs in plants is also known as *vegetative reproduction*. A number of plant species reproduce by this method. Such plants may reproduce asexually by producing buds on leaves, sending out runners from stems, or producing new structures on roots. For centuries, horticulturists have used vegetative reproduction for the cultivation of a number of crops by rooting cuttings of a plant or grafting tissue from one plant to another.

The research of Paul Berg, Stanley Cohen, Herbert Boyer, and their colleagues in the early 1970s made possible another type of cloning, *molecular cloning*. By using recombinant DNA technology, it is now possible to produce clones of whole molecules or fragments of molecules. The molecular segment to be cloned is inserted into a plasmid, which is then inserted into a bacterium. As the bacterium reproduces and carries out its normal life functions, it expresses not only its own DNA, but also any information stored in the DNA fragment that has been added to its own genome. If the fragment added is a gene (as it almost always is), the bacterium produces the protein for which the gene codes. The natural cloning of the original bacterial cell results in countless numbers of daughters, all of which are identical to the original cell, and all of which produce the protein for which the mother cell was engineered. This procedure has been employed in the production of transgenic plants and animals and in pharming, discussed earlier in this chapter, resulting in the availability of a host of new agricultural, pharmaceutical, and industrial products.

Therapeutic and Reproductive Cloning

Two other forms of cloning, *therapeutic cloning* and *reproductive cloning*, have become the focus of intense debate among politicians, scientists, theologians, and the general public. The two types of cloning have very different objectives. The purpose of therapeutic cloning is to generate and then harvest specific cells that can be used in treating medical disorders. The goal of reproductive cloning, on the other hand, is to produce an entirely new organism. Both types of cloning make use of a common procedure, *somatic cell nuclear transfer* (SCNT). The first step in SCNT is to remove the nucleus of an ovum (egg cell) which, in a sense, makes the enucleated (without a nucleus) cell a factory ready to make proteins, without any directions (DNA) as to what proteins to make. Next, the nucleus of a somatic cell is removed and inserted into the enucleated ovum, a process known as *transfection*. The cell factory now has directions as to the products (proteins) it is to make. The hybrid cell is then stimulated by some means to begin dividing. As it divides, it produces clones, all of which are genetically identical to the somatic cell from which the nucleus was originally taken. If that process is allowed to proceed without outside control, a new organism develops that is genetically identical to the one from which the somatic DNA originally came. If that somatic cell came from a frog, the animal that develops will be a frog; if a sheep, it will be a sheep; if a human, it will be a human. Somatic cell nuclear transfer used for this procedure, thus, is a method of reproductive cloning, since it results in the formation of a complete adult organism identical to the parent organism from which the transferred nucleus was taken.

To say that SCNT results in reproductive cloning is not precisely true. While the vast amount of genetic information in a eukaryotic organism is stored in nuclear DNA (DNA in the nucleus of its cells), some small amount of genetic information is also found in mitochondrial DNA. Mitochondrial DNA (mDNA) is DNA found in the mitochondria located in the cytoplasm of a cell. It is closely related to ancestral prokaryotes from which eukaryotes originally evolved. In an SCNT procedure, the vast amount of the genetic information contained in clones will be that from the nuclear DNA transferred from the somatic cell donor. That, of course, explains why adult clones produced by SCNT procedures look so much like the parent organism. But some small amount of genetic information in the clones derives from mitochondrial DNA residing in the original enucleated ovum. One of the fundamental questions still unanswered in SCNT research is how nuclear DNA from the somatic cell donor and mitochondrial DNA from the host egg interact, and what gross effects such interactions will have on the health and well-being of clones produced by SCNT (Committee on Science, Engineering, and Public Policy. Board on Life Sciences 2002, 47–48).

Therapeutic cloning very much resembles reproductive cloning in that DNA from a somatic donor cell is transfected into an enucleated host ovum. The host cell then begins to replicate until it has reached the blastocyst stage of development. A *blastocyst* is an early stage embryo that develops in humans about four to five days after fertilization of the ovum. It consists of fewer than 150 cells. The portion of the blastocyst of interest to scientists is a group of cells on the inner wall of the blastocyst known as *inner cell mass*. These cells are known as *embryonic stem cells*. Embryonic stem cells are of interest to researchers because of two properties: (1) under the right conditions, they are capable of reproducing as undifferentiated cells virtually forever; and (2) under other conditions, they can be stimulated to differentiate into any one of the more than 200 specialized cells (blood cells, nerve cells, muscle cells, etc.) present in the human body.

Controversies about Cloning

Some of the most robust debates over DNA technology center on the use of cloning technologies. Disagreements over the use of molecular cloning for the production of transgenic plants and animals have been discussed earlier. At least as controversial has been the use of somatic cell nuclear transfer for therapeutic and reproductive cloning. This debate has been somewhat clouded by the confusion between therapeutic and reproductive cloning, particularly when applied to humans. Probably the most common generalization that can be made is that the vast majority of individuals and organizations oppose the reproductive cloning of humans, even though those same organizations and individuals may then have very different views about therapeutic cloning.

Human Reproductive Cloning

That having been said, some arguments have been put forward in favor of human reproductive cloning. For example, the technology provides a way for women and couples who are otherwise unable to have children by conventional means to bear a child. Such individuals would include the relatively small number of men who do not produce sperm and women who do not produce eggs, as well as gay men who choose not to have children with genes from a female and lesbians who choose not to have a male sperm donor (Strong 2005). Some proponents of reproductive cloning have also suggested that the technology provides a method for replacing a dead child very much missed by its parents (Connor 2008). Whether one agrees with these arguments or regards them as absurd or outrageous, many advocates of human reproductive cloning argue that the fundamental point may be, after all, that cloning is a technology that individuals should have the right to choose or not, as their own conscience dictates. (See, for example, Pearson 2006.)

Some observers have gone even further and claimed that reproductive cloning might actually be preferable to conventional means of reproduction. One of the world's leading bioethicists, Joseph Fletcher, has written that cloning would not be a dehumanizing activity, as claimed by its opponents, but a good choice from the standpoint of both individuals involved and the general society. At one point, he has argued that

> Good reasons in general for cloning are that it avoids genetic diseases, bypasses sterility, predetermines an individual's gender, and preserves family likenesses. It wastes time to argue over whether we should do it or not; the real moral question is when and why. (Fletcher 1974, 154)

Most people do not find the arguments in favor of reproductive cloning to be persuasive. Many polls in the United States, Europe, Australia, and other countries of the world have found large majorities of respondents in opposition to the practice ("CGS Summary of Public Opinion Polls" 2014). One of the most recent polls, conducted by the Gallup company in 2013, found that 83 percent of respondents believed that reproductive cloning of humans was "morally wrong" and that 60 percent also opposed reproductive cloning of non-human animals ("America: Human and Animal Cloning, ESCs" 2013). The approval/disapproval margin in this poll was roughly comparable to that found in similar polls over the past 25 years ("CGS Summary of Public Opinion Polls" 2014).

When individuals are asked their view about human reproductive cloning, they often respond with an off-the-cuff emotional response that one observer has described as "a group of responses based on yuk, it's unnatural, it's against one's conscience, it's intuitively repellent, it's playing God, it's hubris" (Gillon 1999, 3–4). Although clearly heartfelt and sincere, this response is perhaps difficult to assess rationally. As the writer quoted here observes,

> The trouble is that these gut responses may be morally admirable, but they may also be morally wrong, even morally atrocious, and on their own such gut responses do not enable us to distinguish the admirable from the atrocious. (Gillon 1999, 4)

Other arguments are, perhaps, easier to consider and assess. For example, some critics of human reproductive cloning say that the practice diminishes the value of the individuality and uniqueness of the clone. A person born as a result of this process might be thought of as a "manufactured product" rather than a special human being, deserving of his or her own unique value. Part of the problem might also be that the cloned person would forever live in the shadow of the person from whom he or she was cloned, further degrading the clone's individuality. Finally, using human reproductive cloning for some possibly valid reasons might be another example of taking a step down that "slippery slope" in which the technology might then be used in the future for many less salubrious applications (Center for Genetics and Society 2006a).

At the moment, the arguments for and against human reproductive cloning are all somewhat theoretical. The problem is that current technology has not yet developed to a point where the procedure is safe enough to think of using it with humans, even on an experimental basis. In 2002, a special panel on Scientific and Medical Aspects of Human Cloning of the National Academy of Sciences published a report on the scientific and medical aspects of human reproductive cloning. The committee did not consider ethical and moral questions in its deliberations, simply what the current status of research in the field was. It concluded that the technology was still in a very primitive stage and was not safe to use with humans, even on an experimental basis. One of its findings was that

> Data on the reproductive cloning of animals through the use of nuclear transplantation technology demonstrate that only a small percentage of attempts are successful; that many of the clones die during gestation, even at late stages; that newborn clones are often abnormal or die; and that the procedures may carry serious risks for the mother. In addition, because of the large number of eggs needed for such experiments, many more women would be exposed to the risks inherent in egg donation for a single cloning attempt than for the reproduction of a child by the presently used in vitro fertilization (IVF) techniques. These medical and scientific findings lead us to conclude that the procedures are now unsafe for humans. (Committee on Science, Engineering, and Public Policy. Board on Life Sciences 2002, 93, Appendix B)

Yet, the era of human cloning may not be so far off after all. In 2015, the Chinese company Boyalife announced the imminent opening of a cloning facility in Tianjin capable of producing a million cloned cows by the year 2020. The company noted that the plan would have the capacity also for cloning other specialized mammals, such as racehorses, dogs bred for specialized purposes, such as drug sniffing, and, potentially, human. The technology for cloning humans would be very similar to that for other mammals, the company explained, but it had no immediate plan for cloning human "Frankensteins." The problem, the company said, was no longer technology, but concerns about public reactions to this line of research (Davis 2015). One can hardly refrain from speculating, however, about the future of that technology. (For some thoughts about the future of human cloning, see Seedhouse 2014.)

Therapeutic Cloning

In contrast to public opinion about human reproductive cloning, opinions about therapeutic cloning among the general public are more closely divided. Virginia Commonwealth University (VCU) has conducted a series of polls over the past 15 years of public attitudes toward therapeutic cloning. In the first of these polls, researchers asked interviewees what they thought about the use of cloning technology for medical purposes. They found that 45 percent of respondents approved of such research, while 51 percent opposed the research. Those numbers have remained relatively stable over the history of the polls, rising above 50 percent approval only recently, to 52 percent approval and 47 percent disapproval in 2008 and to 55 percent approval and 40 percent disapproval in 2010 ("CGS Summary of Public Opinion Polls" 2014).

The obvious argument in favor of therapeutic cloning is that it provides a new and powerful tool by which medical science can deal with a variety of human diseases and disorders. Defective or missing cells and organs could be replaced by specially designed engineered products that could be supplied to a patient. Such cells, tissues, and organs could also be much less likely to experience rejection by the body's immune system because they consist of the body's own cells and tissues. Supporters of therapeutic cloning also point to the research value of the technology, providing a new approach to the understanding of human health and disease. Finally, in response to the most common objection to therapeutic cloning, the destruction of embryonic stem cells, many people believe that an early stage embryo has a different moral value than does a later stage embryo, a fetus, or a newborn child. The loss of early stage embryos, therefore, in view of their potential medical value, is well justified.

For opponents of therapeutic cloning, the primary concern usually is that the early stage embryo has the potential to become a living human being. There is no sharp line of demarcation between "not-human" and "human," they say, only a slow progression from primitive human to a living human. It makes no sense, critics say, to destroy a life (that of the early stage embryo) in order to save another life (that of someone with a life-threatening disease). Each life is equally valuable.

Also of considerable importance is the hope that a number of alternatives to the use of embryonic stem cells for therapeutic cloning now appear to be on the horizon, so the use of that technology may no longer be justified at all. Critics of using embryonic stem cells point in particular to an important breakthrough that occurred in 2007. Two research teams, one led by Shinya Yamanaka of Kyoto University, and the other by Junying Yu, working at the University of Wisconsin-Madison, found a way to convert skin cells into cells with the same characteristics as those of embryonic skin cells, a discovery for which Yamanaka received a share of the 2012 Nobel Prize in Physiology or Medicine. If the challenging technical problems that remain with this technology can be solved, a new mechanism for therapeutic cloning may be available that does not require that cells with the potential for life must be sacrificed. (For an excellent review of this issue, see Kfoury 2007.)

Regulations in the United States

Since 1997, the U.S. Congress has made a number of attempts to pass legislation dealing with reproductive cloning. In some cases, those bills failed to pass either house, while in other cases, a bill passed the House of Representatives, but failed in the Senate. The primary reason for this long and tortured history appears to have been an inability to distinguish reproductive from therapeutic cloning. A bill that would ban reproductive cloning alone would, based on previous experience, appear to have a good chance of passing both houses and being signed by a president. But enough people in one or both chambers have felt strongly enough that therapeutic cloning should also be banned, that all bills also included that provision, which made them unacceptable to a majority of the Senate.

The most recent attempt to get such a bill through the Congress was the so-called Human Cloning Prohibition Act of 2105 (H.R. 3498). (As a result of a clerical error, the bill is known by this name although it clearly should have read "2015" rather than "2105.") That bill was introduced into the House by Representative Andy Harris (R-MD) with one co-sponsor. The bill was referred to the Subcommittee on Crime, Terrorism, Homeland Security, and Investigations, but no action was ever taken on the bill (H.R.3498—Human Cloning Prohibition Act of 2105 [*sic*] 2016).

By contrast, a number of states have adopted some type of legislation banning or regulating cloning. The first such act was adopted by the California legislature in 1997. It banned reproductive cloning or cloning for the purpose of inducing artificial pregnancy, but it permitted cloning for research purposes.

As of 2016, seven states—Arizona, Arkansas, Michigan, North Dakota, Oklahoma, South Dakota, and Virginia—have laws that specifically ban both reproductive and therapeutic cloning. Ten other states—California, Connecticut, Illinois, Iowa, Maryland, Massachusetts, Missouri, Montana, New Jersey, and Rhode Island—have more lenient laws that permit therapeutic cloning, but specifically ban reproductive cloning. A number of other states have laws that limit therapeutic cloning in one way or another (typically by banning the use of public funds for the technology), but specifically ban reproductive cloning ("Special Issue: The Threat of Human Cloning Ethics, Recent Developments, and the Case for Action" 2015. Appendix).

Cloning Issues Worldwide

Opposition to reproductive cloning appears to be relatively widespread throughout the world. In a 2007 survey of 193 nations around the world, researchers for the Center for Genetics and Society found that 56 countries (29 percent) had passed specific legislation banning human reproductive cloning, while 137 nations (71 percent) had not yet adopted any such laws. The area in which such legislation was most likely to be found was Western Europe, where 17 nations (77 percent) had adopted laws banning human reproductive cloning, while only five—Andorra, Liechtenstein, Malta, Monaco, and San Marino—had not yet done so. In the Western Hemisphere, 10 nations (40 percent) had laws prohibiting reproductive cloning, while 25 (60 percent) did not. Africa and the Middle East have had by far the least interest in adopting formal legislation on reproductive cloning, with only South Africa, Tunisia, and Israel, of 76 nations in the area, having had adopted any formal legislation on the issue ("National Polices on Human Genetic Modification: A Preliminary Survey" 2007). Other studies show relatively comparable data, but no more recent information appears to be available on the subject. There appears to be no reason to suspect, however, that human cloning is legally encouraged or allowed in any nation of the world today (Matthews 2010; Wheat and Matthews n.d.; probably the most complete current list of cloning laws worldwide is the "International Compilation of Human Research Standards" 2016).

Regional and international agencies have also taken action on reproductive cloning, making recommendations for legislation within a region or providing general standards for the behavior of member states on the issue. One of the most important of these documents is the "Additional Protocol to the Convention for the Protection of Human Rights and Dignity of the Human Being with regard to the Application of Biology and Medicine, on the Prohibition of Cloning Human Beings," adopted by the Council of Europe in December 1998. The convention prohibits "any intervention seeking to create a human being genetically identical to another human being, whether living or dead." It also prohibits any exception to this ban, even in the case of a completely sterile couple ("Details of Treaty No. 168" 1998, Article 1, Section 2). Framers of the document made clear that the purpose of the convention was to go beyond the specific issue of reproductive cloning and to "enshrine important principles which provide the ethical basis for further biological and medical developments, both now and in the future" ("Details of Treaty No. 168" 1998). As of mid-2016, 23 nations had ratified the treaty and 10 nations had signed, but not ratified the treaty. It came into force on January 3, 2001, when five nations had signed the treaty.

Perhaps the most fascinating statement on human reproductive cloning has been what has transpired with the United Nations. In 2000, France and Germany proposed that the United Nations convene a conference to write a convention that would ban human reproductive cloning, but that would leave therapeutic cloning for a later date. Objections to this proposal were raised by other nations, particularly Spain and the United States (as well as the Vatican City, acting as an observer). These countries lobbied for a more comprehensive treatment that would ban all kinds of cloning, both reproductive and therapeutic. Negotiations on the treaty dragged on for four years until a vote was held on March 8, 2005. The proposed draft of the treaty, banning all forms of cloning, was adopted by a vote of 84 in favor, 34 opposed, with 37 abstentions, and 33 nations absent from the vote. In the end, the vote had little significance not only because the final treaty consisted only of recommendations and was not binding on UN members, but also because the vote was so inconclusive (it passed with 45 percent of the total vote) ("The United Nations Human Cloning Treaty Debate, 2000–2005" 2006).

The majority of countries that have specific laws about cloning restrict both reproductive and therapeutic cloning. In one of the most recent surveys of such laws, 36 countries worldwide fell into that category. By contrast, a relatively small number of countries specifically allowed therapeutic cloning, even though they banned reproductive cloning. Those countries were Belgium, China, Colombia, Finland, Israel, Japan, New Zealand, Singapore, South Korea, Sweden, Thailand, Turkey, the United Kingdom, and the United States (Matthews 2010).

Public opinion about reproductive and therapeutic cloning worldwide appears to reflect attitudes in the United States to a considerable extent. In general, Europeans tend to be supportive of therapeutic cloning, provided that it is regulated. In a 2003 EOS Gallup Europe poll, 55 percent of the 15 EU members favored the use of therapeutic cloning, while 43 opposed the practice. The highest level of approval among the 15 EU members came from Spain, with 79 percent approval compared to 18 percent disapproval, followed by Portugal (72 percent/22 percent), and Italy (65 percent/33 percent). The greatest level of disapproval was expressed in Greece (33 percent/65 percent), Ireland (37 percent/59 percent), and Austria (39 percent/54 percent). Among the 13 candidate countries at the time, approval for therapeutic cloning was somewhat lower than it was for the 15 EU members, 44 percent approval overall compared to 47 percent disapproval.

Public opinion polls about therapeutic cloning have also been conducted in a number of nations outside the European Union. Such a poll in Australia in 2006, for example, found that 58 percent of respondents supported therapeutic cloning (30 percent "strongly support" and 28 percent "somewhat support"), while 18 percent were totally opposed (10 percent "strongly opposed" and 8 percent "somewhat opposed") (Research Australia 2006, 13). A poll of Canadian public opinion in 2003 produced generally similar results, with 53 percent of respondents indicating that they supported therapeutic cloning overall and 32 percent opposing the practice. Perhaps the most interesting trend within this poll was the finding that a large majority of respondents between the ages of 18 and 24 (66 percent) supported therapeutic cloning (Leger Marketing 2003, 3).

The Double-Edged Sword of CRISPR Technology

The discovery of new gene-editing technologies mentioned in Chapter 1, such as CRISPR-Cas9, has opened the door to a number of new research and practical applications that have significant benefits and risks for society. For example, they make it possible for researchers to design and create new cellular and animal models for diseases that can be studied in the laboratory more quickly and efficiently than has been possible before. Such models should be able to provide answers to the etiology, cure, prevention, and treatment of diseases that have long been recalcitrant to other types of medical research.

Gene editing should also be able to increase the efficiency of traditional approaches to genetic engineering, creating new routes to the production of synthetic pharmaceuticals, fuels, agricultural products, and other commercial materials. It has the potential also for new forms of gene surgery, that is, the removal of alteration of genes implicated in the development of genetic diseases and disorders.

One of the most profound applications of gene editing may be in the field known as *gene drive*, sometimes suggested as a mechanism for completely eliminating certain types of organisms with largely negative effects on humans, other animals, and certain plant species. Gene drive is a mechanism by which some of the most fundamental principles of genetics are avoided or re-directed. For example, the laws of genetics predict that two alleles (versions) of a gene have an equal chance of being transmitted to further generations. In rare circumstances, however, an additional factor may be present in an organism that increases the probability that one of the two alleles of a gene will be transmitted. In such a case, that allele can become more and more predominant in the organism, resulting in the elimination of the other allele. (For a visual representation of this event, see "FAQs: Gene Drives" 2016.) The availability of gene-editing technology makes it possible for researchers to initiate or promote this process in an organism, making it possible to eliminate a particular variation or species of an organism.

This form of research has significant value for dealing with certain health issues now facing the human race. For example, the possibility of completely eliminating the anopheles mosquito responsible for the transmission of malaria is very appealing to some public health workers. In one broad sweep, one of the world's most treacherous diseases could be eliminated, or at least brought under control.

Many ecologists point out the risks involved in this type of research, however. They note that gene-editing techniques are relatively new inventions and that researchers are not entirely clear as to how they might affect entire ecosystems. In June 2016, a panel of the National Academies of Science, Engineering, and Medicine formally endorsed continued research on gene drives. It took note of the risks associated with the technology, however, and recommended that research move forward only very slowly and carefully making use of "carefully controlled field trials" (Harmon 2016).

Conclusion

DNA technology has made possible a dazzling array of new procedures and products to aid in fighting crime, to augment and improve the world's food supply, to help medical sciences have a better understanding of the nature of genetic illnesses and to provide pharmaceuticals to use against those and other kinds of disease, to find ways of curing or even eliminating some genetic disorders, and to produce exact copies of animals that can then be used in a host of beneficial ways. None of these technologies has been quickly and easily adopted, however, as each brings with it certain risks to human health and/or the environment and, in many cases, raises profound social, political, and ethical questions. As is usually the case with progress in science, the great challenge is to find ways of gaining the greatest benefit of each new technology while reducing to the greatest extent possible any serious risks posed to the natural and human world.

References

"The Acceptance of Therapeutic and Reproductive Cloning by the Public." 2003. EOS Gallup Europe. http://www.geneticsandsociety.org/downloads/200301_EOS_Gallup_Europe.pdf. Accessed on April 12, 2016.

"America: Human and Animal Cloning, ESCs." 2013. Gallup. http://www.geneticsandsociety.org/article.php?id=7480. Accessed on April 9, 2016.

American Board of Genetic Counseling. 2016. http://www.abgc.net/ABGC/AmericanBoardofGeneticCounselors.asp. Accessed on April 8, 2016.

Andrews, Lori B., et al., eds. 1994. *Assessing Genetic Risks*. Washington, DC: National Academy Press.

"AquAdvantage Salmon." 2016. U.S. Food and Drug Administration. http://www.fda.gov/AnimalVeterinary/DevelopmentApprovalProcess/GeneticEngineering/GeneticallyEngineeredAnimals/ucm280853.htm. Accessed on April 6, 2016.

Aronson, Jay D. 2007. *Genetic Witness: Science, Law, and Controversy in the Making of DNA Profiling.* New Brunswick, NJ: Rutgers University Press.

Barlett, Donald L., and James B. Steele. 2008. "Monsanto's Harvest of Fear." *Vanity Fair.* http://www.vanityfair.com/politics/features/2008/05/monsanto200805. Accessed on April 5, 2016.

Belson, Neil A. 2000. "US Regulation of Agricultural Biotechnology: An Overview." *AgBioForum.* 3(4): 268–280. http://www.agbioforum.org. Accessed on April 7, 2016.

Bergthorsson, Ulfar, et al. 2003. "Widespread Horizontal Transfer of Mitochondrial Genes in Flowering Plants." *Nature.* 424(6945): 197–201.

Blackmore, Willy. 2016. "General Mills Will Label GMOs on Products Nationwide." The Cornucopia Institute. http://www.cornucopia.org/2016/03/general-mills-will-label-gmos-on-products-nationwide/. Accessed on April 6, 2016.

Bodnar, Anastasia. 2010. "GMOs Could Render Important Antibiotics Worthless." Science 2.0. http://www.science20.com/genetic_maize/blog/gmos_could_render_important_antibiotics_worthless_1. Accessed on April 5, 2016.

"BRCA1 and BRCA2: Cancer Risk and Genetic Testing." 2015. National Cancer Institute. http://www.cancer.gov/about-cancer/causes-prevention/genetics/brca-fact-sheet#q1. Accessed on April 8, 2016.

Bryan, Jenny, and John Clare. 2001. "Primates as Recipients: Animal Arguments: Pigs and People." Frontline. http://www.pbs.org/wgbh/pages/frontline/shows/organfarm/rights/primates.html. Accessed on April 7, 2016.

"Bt Corn Pollen and Monarch Butterflies" (six articles). 2001. *Proceedings of the National Academy of Sciences of the United States of America.* 98(21): 11908–11942.

Caruso, Denise. 2004. "New 'Future of Food' Movie Is a Hit!" Organic Consumers Association. http://www .organicconsumers.org/ge/future-of-food.cfm. Accessed on April 5, 2016.

"Cellular and Gene Therapy Products." 2015. U.S. Food and Drug Administration. http://www.fda.gov/ biologicsbloodvaccines/cellulargenetherapyproducts/ default.htm. Accessed on April 9, 2016.

"CGS Summary of Public Opinion Polls." 2014. http:// www.geneticsandsociety.org/article.php?id=401#reprodata. Accessed on April 9, 2016.

"CODIS Brochure." 2016. U.S. Federal Bureau of Investigation. https://www.fbi.gov/about-us/lab/biometric-analysis/codis/codis_brochure. Accessed on April 5, 2016.

"CODIS—NDIS Statistics." 2016. U.S. Federal Bureau of Investigation. https://www.fbi.gov/about-us/lab/biometric-analysis/codis/ndis-statistics. Accessed on April 5, 2016.

Committee on Science, Engineering, and Public Policy. Board on Life Sciences. 2002. *Scientific and Medical Aspects of Human Reproductive Cloning*. Washington, DC: National Academies Press.

"Congressional Train Rolling to Pre-Empt States on GMO Labels." 2016. Food Safety News. http://www .foodsafetynews.com/2016/03/congressional-train-is-moving-to-pre-empt-states-on-gmo-labeling/#.VwWH4fkrJD8. Accessed on April 6, 2016.

Connor, Steve. 2008. "'Now We Have the Technology That Can Make a Cloned Child'." *The Independent*. http://www.independent.co.uk/news/science/ now-we-have-thetechnology-that-can-make-a-cloned-child-808625.html. Accessed on April 9, 2016.

"Coordinated Framework for Regulation of Biotechnology." 1986 Office of Science and Technology Policy. https:// www.aphis.usda.gov/brs/fedregister/coordinated_ framework.pdf. Accessed on April 7, 2016.

"Cosmetic Surgery National Data Bank Statistics." 2015. American Society for Aesthetic Plastic Surgery. http://www .surgery.org/sites/default/files/Stats2015.pdf. Accessed on April 8, 2016.

"Country Reports: GMOs in EU Member States." 2016. GMO Compass. http://www.gmo-compass.org/eng/news/ country_reports/. Accessed on April 11, 2016.

Cummins, Joe. 2014. "Risks of Edible Transgenic Vaccines." Engineered Ignorance. https://engineeredignorance. wordpress.com/2013/03/09/old-news-on-edible- transgenic-vaccines-and-food/. Accessed on April 7, 2016.

Davis, Rebecca. 2015. "China 'Clone Factory' Scientist Eyes Human Replication." *Yahoo! News.* https://www .yahoo.com/news/china-clone-factory-scientist-eyes- human-replication-061141389.html. Accessed on April 10, 2016.

"Details of Treaty No. 168." 1998. Council of Europe. http://www.coe.int/en/web/conventions/full-list/-/ conventions/treaty/168. Accessed on April 12, 2016.

DeWeerdt, Sarah. 2014. "Gene Therapy: A Treatment Coming of Age." *The Pharmaceutical Journal.* http://www .pharmaceutical-journal.com/news-and-analysis/feature/ gene-therapy-a-treatment-coming-of-age/20066677 .article. Accessed on April 8, 2016.

"Directive 2001/18/ec of the European Parliament and of the Council of 12 March 2001." 2001. *Official Journal of the European Community.* April 17, 2001, L106/1–L106/38.

"Eat Up Your Vaccines." 2000. Grain. https://www.grain .org/article/entries/245-eat-up-your-vaccines. Accessed on April 7, 2016.

"EU Alters GMO Assessments to Satisfy Resistant Member States." 2006. National Health Federation. http://www .thenhf.com/eu-alters-gmo-assessments-to-satisfy-resistant- member-states/. Accessed on April 11, 2016.

"Familial Adenomatous Polyposis." 2013. Genetics Home Reference. https://ghr.nlm.nih.gov/condition/familial-adenomatous-polyposis. Accessed on April 8, 2016.

"FAQs: Gene Drives." 2016. Wyss Institute. Harvard University. http://wyss.harvard.edu/staticfiles/newsroom/pressreleases/Gene%20drives%20FAQ%20FINAL.pdf. Accessed on June 15, 2016.

Fernandez-Cornejo, Jorge, et al. 2014. "Genetically Engineered Crops in the United States." Economic Research Service. *Economic Research Report Number 162*. http://www.ers.usda.gov/media/1282246/err162.pdf. Accessed on April 5, 2016.

Fletcher, Meg. 2007. "A Question of Privacy." Business Insurance. http://www.businessinsurance.com/article/20070318/FOCUS/100021346. Accessed on April 8, 2016.

Fletcher, Joseph. 1974. *The Ethics of Genetic Control: Ending Reproductive Roulette*. New York: Anchor Books.

Food and Agriculture Organization. United Nations. 2004. "The State of Food and Agriculture 2003–2004: Agricultural Biotechnology. Meeting the Needs of the Poor?" Rome: Food and Agriculture Organization.

"Forensics." 2016. Interpol. http://www.interpol.int/INTERPOL-expertise/Forensics/DNA. Accessed on April 11, 2016.

Fox, Jeffrey. 1998. "Germline Gene Theory Contemplated." *Nature Biotechnology*. 16(5): 407.

Friedlander, Blaine P., Jr. 1999. "Toxic Pollen from Widely Planted, Genetically Modified Corn Can Kill Monarch Butterflies, Cornell Study Shows." *Cornell News*, May 19. Available online. http://www.news.cornell.edu/stories/1999/04/toxic-pollen-bt-corn-can-kill-monarch-butterflies. Accessed on April 5, 2016.

Funk, Cary, and Lee Rainie. 2015. "Public Opinion about Food." Pew Research Center. Internet, Science, and Tech. http://www.pewinternet.org/2015/07/01/chapter-6-public-opinion-about-food/. Accessed on April 6, 2016.

"Gene Therapy." 2016. U.S. Food and Drug Administration. http://google2.fda.gov/search?q=gene+therapy&client=FDAgov&site=FDAgov&lr=&proxystylesheet=FDAgov&requiredfields=-archive%3AYes&output=xml_no_dtd&getfields=*. Accessed on April 9, 2016.

"Genetic Engineering." 2009. Responsa in a Moment. http://www.responsafortoday.com/moment/3_1.htm. Accessed on April 9, 2016.

"Genetic Privacy Laws." 2008. National Conference of State Legislatures. http://www.ncsl.org/research/health/genetic-privacy-laws.aspx. Accessed on April 8, 2016.

"Genetically Engineered Foods May Cause Rising Food Allergies (in two parts). 2007. Institute for Responsible Technology. http://responsibletechnology.org/genetically-engineered-foods-may-cause-rising-food-allergies-part-one/ and http://responsibletechnology.org/genetically-engineered-foods-may-cause-rising-food-allergies-part-two/. Accessed on April 5, 2016.

Gillon, Raanan. 1999. "Human Reproductive Cloning—A Look at the Arguments against It and a Rejection of Most of Them." *Journal of the Royal Society of Medicine.* 92 (1): 3–12.

"GMO Legislation." 2016. European Commission. http://ec.europa.eu/food/plant/gmo/legislation/index_en.htm. Accessed on April 11, 2016.

Goubau, Alain. 2011. "The AquAdvantage Salmon Controversy—a Tale of Aquaculture, Genetically Engineered Fish and Regulatory Uncertainty."

https://dash.harvard.edu/bitstream/handle/1/8789564/
Goubau_Alain-_Food_%26amp__Drug_Law_-_
Final_Paper_-_AquAdvantage_Salmon%5B1%5D.pdf?
sequence=1. Accessed on April 6, 2016.

Gruber, Jeremy. 2009. "The New Genetic Information
Nondiscrimination Act: How It Came to Pass and
What It Does." GeneWatch. http://www.councilfor
responsiblegenetics.org/GeneWatch/GeneWatchPage
.aspx?pageId=185. Accessed on April 8, 2016.

"Guide to U.S. Regulation of Genetically Modified Food
and Agricultural Biotechnology Products." 2001. Pew
Initiative on Food and Technology. http://www.pewtrusts
.org/~/media/legacy/uploadedfiles/wwwpewtrustsorg/
reports/food_and_biotechnology/hhsbiotech0901pdf.pdf.
Accessed on April 6, 2016.

Harmon, Amy. 2016. "Panel Endorses 'Gene Drive'
Technology That Can Alter Entire Species." *New York
Times*. http://www.nytimes.com/2016/06/09/science/
national-academies-sciences-gene-drive-technology.html?_
r=0. Accessed on June 15, 2016.

Harris, J., and S. Chan. 2008. "Enhancement Is Good for
You!: Understanding the Ethics of Genetic Enhancement."
Gene Therapy. 15(1): 338–339.

Harrison, Pete. 2009. "EU Upholds Austria, Hungary Right
to Ban GM Crops." http://www.reuters.com/article/
us-eu-gmo-bans-idUSTRE5212OL20090302. Accessed on
April 11, 2016.

"The Health Risks of GM Foods: Summary and Debate."
2016. Seeds of Deception. http://globalsuccessleaders.com/
pdfs/The_Health_Risks_of_GM_Foods_52_studies.pdf.
Accessed on April 5, 2016.

"How FDA Regulates Food from Genetically Engineered
Plants." 2015. U.S. Food and Drug Administration. http://
www.fda.gov/Food/FoodScienceResearch/GEPlants/
ucm461831.htm/ Accessed on April 6, 2016.

"H.R.3498—Human Cloning Prohibition Act of 2105 [*sic*]." 2016. Congress.Gov. https://www.congress.gov/ bill/114th-congress/house-bill/3498/text. Accessed on April 10, 2016.

"H.R. 5269 (109th): Genetically Engineered Food Right to Know Act." 2016. GovTrack.US. https://www.govtrack .us/congress/bills/109/hr5269. Accessed on April 6, 2016.

"International Compilation of Human Research Standards." 2016. U.S. Department of Health and Human Services. Available online at http://www.hhs.gov/ohrp/international/ index.html. Accessed on April 12, 2016.

International DNA Databases. 2016. Gene Watch UK. http://www.genewatch.org/sub-566699. Accessed on April 11, 2016.

"Introduction and Summary of Findings." 2011. Council for Responsible Genetics. http://www.councilforrespon siblegenetics.org/dnadata/exec.html. Accessed on April 11, 2016.

"Judge Dwyer Issues Written Decision in Landmark DNA Case Won by Legal Aid's DNA Unit." 2016. The Legal Aid Society. http://www.legal-aid.org/ en/mediaandpublicinformation/inthenews/judge dwyerissueswrittendecisioninlandmarkdnacasewonby legalaid%E2%80%99sdnaunit.aspx. Accessed on April 5, 2016.

"Judgment of the Court 21–03–2000." 2000. http://www .asser.nl/upload/eel-webroot/www/cases/HvJEG/699j0006 .htm. Accessed on April 11, 2016.

Katz, Gregory, and Stuart O. Schweitzer. 2010. "Implications of Genetic Testing for Health Policy." *Yale Journal of Health Policy, Law, and Ethics*. 10(1): 90–134.

Kayser, Oliver, and Heribert Warzecha, eds. 2012. *Pharmaceutical Biotechnology: Drug Discovery and Clinical Applications*. Weinheim, Germany: Wiley-Blackwell.

Keeling, Patrick J., and Jeffrey D. Palmer. 2008. "Horizontal Gene Transfer in Eukaryotic Evolution." *Nature Reviews Genetics.* 9(8): 605–618.

Keim, Brandon. 2008. "Genetic Protections Skimp on Privacy, Says Gene Tester." Wired Science. http://www .wired.com/2008/05/genetic-protect/. Accessed on April 8, 2016.

Kfoury, Charlotte. 2007. "Therapeutic Cloning: Promises and Issues." *Mcgill Journal of Medicine.* 10(2): 112–120.

King, David. 1997. "Social Issues Raised by Gene Therapy." http://www.hgalert.org/topics/hge/geneTherapy.htm. Accessed on April 8, 2016.

Kirk, Leigh. 2008. "WalMart Bans rBGH Tainted Milk." Natural News.com. http://www.naturalnews.com/022 913_rBGH_milk_cows.html. Accessed on April 6, 2016.

Lambert, Bruce. 1998. "Giuliani Backs DNA Testing of Newborns for Identification." *New York Times.* http://www.nytimes.com/1998/12/17/nyregion/giuliani-backs-dna-testing-of-newborns-for-identification.html. Accessed on April 5, 2016.

Lander, Eric S., and Bruce Budowle. 1994. "Commentary: DNA Fingerprinting Dispute Laid to Rest." *Nature.* 371(6500): 735–738.

Langridge, William H. R. 2000. "Edible Vaccines." *Scientific American.* 283(3): 66–71.

Leary, Warren E. 1996. "Genetic Engineering of Crops Can Spread Allergies, Study Shows." *New York Times.* http://www.nytimes.com/1996/03/14/us/genetic-engineering-of-crops-can-spread-allergies-study-shows.html. Accessed on April 5, 2016.

Ledford, Heidi. 2013. "Transgenic Salmon Nears Approval." *Nature.* 497(7447): 17–18. http://www.nature.com/news/ transgenic-salmon-nears-approval-1.12903. Accessed on April 6, 2016.

Leger Marketing. 2003. "How Canadians Feel about Cloning." Montréal: Leger Marketing.

"Major Gaps between the Public, Scientists on Key Issues." 2015. Pew Research Center. Internet, Science, and Tech. http://www.pewinternet.org/interactives/public-scientists-opinion-gap/. Accessed on April 6, 2016.

"Making Sense of Your Genes: A Guide to Genetic Counseling." 2008. Chicago: National Society of Genetic Counselors, Inc.; Washington, DC: Genetic Alliance.

Matthews, Kirstin. 2010. "Overview of World Human Cloning Policies." OpenStax. http://cnx.org/contents/4O_mfmxO@1/Overview-of-World-Human-Clonin. Accessed on April 12, 2016.

McGeehan, Patrick. 2007. "Spitzer Wants DNA Sampling in Most Crimes." *New York Times*. http://www.nytimes.com/2007/05/14/nyregion/14dna.html?_r=0. Accessed on April 5, 2016.

McHughen, Alan. 2008. "Labeling Genetically Modified (GM) Foods." http://www.agribiotech.info/more-details-on-specific-issues/files/McHugen-Labeling%20sent%20to%20web%2002.pdf. Accessed on April 6, 2016.

Meade, Harry, and Carol Ziomek. 1998. "Urine as a Substitute for Milk?" *Nature Biotechnology*. 16(1): 21–22.

Meller, Paul, and Andrew Pollack. 2004. "Europeans Appear Ready to Approve a Biotech Corn." *New York Times*. May 15, 2004: C1.

Murphy, Erin E. 2015. *Inside the Cell: The Dark Side of Forensic DNA*. New York: Nation Books.

"National Polices on Human Genetic Modification: A Preliminary Survey." 2007. Center for Genetics and Society. http://www.geneticsandsociety.org/article.php?id=304. Accessed on April 12, 2016.

"Newborn Screening Tests." 2015. ADAM Multimedia Encyclopedia. http://franciscanalliance.adam.com/content.aspx?productId=117&pid=1&gid=007257. Accessed on April 7, 2016.

Papademetriou, Theresa. 2014. "Restrictions on Genetically Modified Organisms: European Union." Library of Congress. https://www.loc.gov/law/help/restrictions-on-gmos/eu.php#_ftn61. Accessed on April 12, 2016.

Pearson, Yvette. 2006. "Never Let Me Clone?: Countering an Ethical Argument against the Reproductive Cloning of Humans." *Science and Society*. 7(7): 657–660.

"Pharm and Industrial Crops." 2003. Union of Concerned Scientists. http://www.ucsusa.org/sites/default/files/legacy/assets/documents/food_and_agriculture/pharm cropsucs403.pdf. Accessed on April 7, 2016.

"Pharming for Farmaceuticals." 2016. Learn.Genetics. http://learn.genetics.utah.edu/content/science/pharming/. Accessed on April 7, 2016.

"Pharming the Field: A Look at the Benefits and Risks of Bioengineering Plants to Produce Pharmaceuticals." 2002. Pew Initiative on Food and Biotechnology. U.S. Food and Drug Administration. Cooperative State Research, Education and Extension Service, U.S. Department of Agriculture. http://www.pewtrusts.org/~/media/legacy/uploadedfiles/wwwpewtrustsorg/reports/food_and_biotechnology/pifbpharmingfieldspdf.pdf. Accessed on April 7, 2016.

"Plant Genes Imported from Unrelated Species More Often Than Previously Thought, IU Biologists Find." 2003. IU News Room. http://newsinfo.iu.edu/news/page/normal/1028.html. Accessed on April 5, 2016.

Pollack, Andrew. 2015. "White House Orders Review of Rules for Genetically Modified Crops." *New York*

Times. http://www.nytimes.com/2015/07/03/business/white-house-orders-review-of-biotechnology-regulations.html. Accessed on April 7, 2016.

"Regulation of Genetic Tests." 2015. National Human Genome Research Institute. https://www.genome.gov/10002335. Accessed on April 8, 2016.

Rehbinder, Eckard, et al. 2008. *Pharming: Promises and Risks of Biopharmaceuticals Derived from Genetically Modified Plants and Animals.* New York: Springer.

"Report on the Food and Drug Administration's Review of the Safety of Recombinant Bovine Somatotropin." 2014. U.S. Food and Drug Administration. http://www.fda.gov/AnimalVeterinary/SafetyHealth/ProductSafetyInformation/ucm130321.htm#IGF-I. Accessed on April 6, 2016.

Research Australia. 2006. "Health & Medical Research Public Opinion Poll." 2006. Sydney: Research Australia Limited.

"Research Cloning Arguments Pro and Con." 2006. Center for Genetics and Society. http://www.geneticsandsociety.org/article.php?id=284. Accessed on April 10, 2016.

"Results of Search in US Patent Collection db for: AN/Monsanto." 2016. USPTO Patent Full-Text and Image Database. http://patft.uspto.gov/netacgi/nph-Parser?Sect1=PTO2&Sect2=HITOFF&u=%2Fnetahtml%2FPTO%2Fsearch-adv.htm&r=0&p=1&f=S&l=50&Query=AN%2FMonsanto%24&d=PTXT. Accessed on April 5, 2016.

Robinson, Clair. 2009. "Genetically Modified Justice?" Cropchoice.com. http://www.cropchoice.com/leadstry562e.html?recid=2135. Accessed on April 5, 2016.

"Rogue GM Plant Warning." 2002. *BBC News.* http://news.bbc.co.uk/1/hi/sci/tech/1800322.stm. Accessed on April 5, 2016.

"Search Results for 'Genetically Modified Foods.'" 2016. The Pew Charitable Trusts. http://www.pewtrusts.org/ en/search?q=genetically%20modified%20foods. Accessed on April 6, 2016.

Seedhouse, Erik. 2014. *Beyond Human: Engineering Our Future Evolution.* Heidelberg: Springer.

Sijmons, Peter C., et al. 1990. "Production of Correctly Processed Human Serum Albumin in Transgenic Plants." *Nature BioTechnology.* 8(3): 217–221.

"Special Issue: The Threat of Human Cloning Ethics, Recent Developments, and the Case for Action." 2015. *The New Atlantis.* 46(Special Issue): 9–146. https://www.jstor.org/ stable/43551612?seq=1#page_scan_tab_contents. Accessed on April 10, 2016.

"State Legislation Addressing Genetically Modified Organisms." 2015. National Conference of State Legislatures. http://www.ncsl.org/research/agriculture-and-rural-development/state-legislation-addressing-genetically-modified-organisms.aspx#ND. Accessed on April 6, 2016.

Strong, Carson. 2005. "Reproductive Cloning Combined with Genetic Modification." *Journal of Medical Ethics.* 31(11): 654–658.

Szebik, Imre, and Kathleen Cranley Glass. 2001. "Ethical Issues of Human Germ-Cell Therapy: A Preparation for Public Discussion." *Academic Medicine.* 76 (1): 32–38.

Takeyama, Natsumi, Hiroshi Kiyono, and Yoshikazu Yuki. 2015. "Plant-Based Vaccines for Animals and Humans: Recent Advances in Technology and Clinical Trials." *Therapeutic Advances in Vaccines.* 3(5–6): 139–154.

Thompson, Larry. 2000. "Human Gene Therapy: Harsh Lessons, High Hopes." *FDA Consumer.* 34(5): 19–24.

"The United Nations Human Cloning Treaty Debate, 2000–2005." 2006. Center for Genetics and Society.

http://geneticsandsociety.org/article.php?id=338&&print safe=1. Accessed on April 12, 2016.

United Network for Organ Sharing. 2016. https://www.unos .org/. Accessed on April 6, 2016.

U.S. Department of Health and Human Services. 1994. "Interim Guidance on the Voluntary Labeling of Milk and Milk Products From Cows That Have Not Been Treated with Recombinant Bovine Somatotropin." *Federal Register.* February 10: 6279.

U.S. Department of Health and Human Services. Food and Drug Administration. 1993. "Application of Current Statutory Authorities to Human Somatic Cell Therapy Products and Gene Therapy Products." *Federal Register.* 58(197): 53248–53251.

U.S. Department of Health and Human Services. Food and Drug Administration. Center for Biologics Evaluation and Research. 1999. *Guidance For Industry. Public Health Issues Posed by the Use of Nonhuman Primate Xenografts in Humans.* Washington, DC: Food and Drug Administration.

U.S. Government Accountability Office. 2006. "Nutrigenetic Testing: Tests Purchased from Four Web Sites Mislead Consumers." GAO-06-977T.

"US Polls on GE Food Labeling." 2016. Center for Food Safety. http://www.centerforfoodsafety.org/issues/976/ ge-food-labeling/us-polls-on-ge-food-labeling. Accessed on April 6, 2016.

"USTR Details Urgency of Ending EU Ban on Biotech Food." 2003. United States Mission to the European Union. http://iipdigital.usembassy.gov/st/english/text trans/2003/05/20030513111221yessedo0.7209436 .html#axzz45Yla2Z7m. Accessed on April 11, 2016.

"Vaccine Development Using Recombinant DNA Technology." 2008. CAST Issue Paper. Number 38.

http://www.cast-science.org/download.cfm?Publication
ID=2937&File=1e30d72d204bb231d4e8261b472ef5a214
f6TR. Accessed on April 7, 2016.

Wang, Yanli, et al. 2013. "Expression Systems and Species
Used for Transgenic Animal Bioreactors." Biomedical
Research International. 2013(2): 1–9. http://www.ncbi
.nlm.nih.gov/pmc/articles/PMC3613084/. Accessed on
April 7, 2016.

Weiser, Benjamin, and Joseph Goldstein. 2016. "Ex-Official
Says Medical Examiner Forced Her Out over DNA
Technique." *New York Times.* http://www.nytimes
.com/2016/02/19/nyregion/ex-official-says-medical-
examiner-forced-her-out-over-dna-technique.html?_r=0.
Accessed on April 6, 2016.

Wheat, Kathryn, and Kirstin Matthews. n.d. http://www
.ruf.rice.edu/~neal/stemcell/World.pdf. Accessed on
April 12, 2016.

Willing, Richard. 2003. "White House Seeks to Expand
DNA Database." *USA Today.* http://www.usatoday
.com/news/washington/2003-04-15-dna-usat_x.htm.
Accessed on April 5, 2016.

Witten, Celia M. 2007. "Human Gene Therapies: Novel
Product Development." *FDA Consumer.* http://www
.fda.gov/downloads/ForConsumers/ConsumerUpdates/
UCM219770.pdf. Accessed on April 9, 2016.

Word, Charlotte. 2010. "What Is LCN?—Definitions and
Challenges." Promega. http://www.promega.com/
resources/profiles-in-dna/2010/what-is-lcn-definitions-
and-challenges/. Accessed on April 5, 2016.

Yu, Natalie. 2008. "Edible Vaccines." *MMG 445 Basic
Biotechnology eJournal.* 4: 104–109.

3   Perspectives

Introduction

Progress in DNA research is exciting, not only because of the technological developments that evolve from such research, but also because of the often significant and sometimes dramatic social changes that may result from it. This chapter provides individuals with an opportunity to express their own views on some aspect of these advances. Some essays describe and discuss progress in DNA technology itself (such as the discovery of the CRISPR technology), while others provide an overview of some of the social, political, economic, and other issues that have developed as a result of that research.

Optogenetics
Arpita Dave

Optogenetics is the combination of genetic and optical methods to activate or inhibit well-defined events in specific cells of living tissue and animals. In optogenetics, a light-sensitive protein codon sequence is engineered into a gene construct, which is then transfected into cells. Optogenetic events are then produced by exposing those cells to light. One important feature

---

As part of a national campaign of public sensitization against CRISPR-Cas9 dangers, a group of activists from Alliance Vita protest in the center of Lyon (Southeast France). The CRISPR-Cas9 technique allows scientists to modify the animal, vegetal, or human DNA sequence. (Serge Mouraret/ Corbis via Getty Images)

of the process is that the light does not affect cells when light sensitive proteins are not present (Deisseroth 2011).

The proteins sensitive to light in prokaryotes and eukaryotes are called *retinal opsins* or *rhodopsins*. Conformational changes that take place in these proteins when exposed to specific wavelengths of light open channels and create electrical signals. Prokaryotes use opsins mainly for light-driven energy generation, whereas eukaryotic opsins present in the retina of the eye allow transmission of impulses to the brain. Unlike eukaryotic rhodopsins, which are coupled to intracellular second messenger cascade that indirectly influence ion channels, prokaryotic rhodopsins directly transduce photons into a current (Deisseroth 2015; Zhang 2011).

Optogenetics has been employed by researchers to control specific neurons as a way of learning about neuronal connections involved in learning, memory, emotions, motor functions, and brain diseases. It could also become a promising tool for non-neuronal cell-specific activation or inhibition of signaling pathways from the scale of cells to organisms with high spatial (microns deep) and temporal (milliseconds) precision (Gorostiza 2008).

In 2002, Austro-English physiologist Gero Miesenböck first engineered three genes encoding retinal opsins of Drosophila (fruit fly) in a rat hippocampal neuronal culture. A year later, George Navgel, Ernst Bamberg, Peter Hegemann, and other researchers discovered an opsin that drives phototaxis (the movement of a body in response to exposure to light) in the green alga *Chlamydomonas reinhardtii* to which they gave the name channelrhodopsin2. They expressed channelrhodopsin2 into oocyte and HEK (human embryonic kidney) cells and demonstrated that it can be used for depolarization of cells. The discovery of this channelrhodopsin2 has been considered revolutionary as the green algal channelrhodopsin2 (ChR2) responded to light more speedily than did opsins from Drosophila (Boyden 2011).

In 2005, Deisseroth and Boyden for the first time expressed algal opsin, channelrhodopsin2 in rat neuronal culture and found that it was possible to selectively stimulate neurons by light without harming other cells. Further they established a method to fire up neurons of the hippocampus in live animals to trace out the neuronal circuits involved in motor function. Guoping Feng, in 2007, established transgenic mice expressing channelrhodopsin2 in particular neurons to study the neuronal circuits (Boyden 2011).

A number of prokaryotic opsins such as channelrhodopsin (Na^+, K^+, Ca^{2+}, H^+ ion channel), bacteriorhodopsin (H^+ ion channel), halorhodopsin (Cl^- ion channels), and synthetic rhodopsins called optoXRs, were then put to use by researchers to come up with opsins that are stimulated and respond to light faster as a way of studying the living cell processes in real time (Deisseroth 2015; Zhang 2011).

Optogenetics require four main components: (1) a gene construct containing a promoter sequence for a specific cell type followed by gene sequences encoding for a light-sensitive protein channel (actuator) and fluorescent protein (reporter), (2) a transfection system, (3) opto fibers to deliver light, and (4) sensors, that is, electrodes capable of sensing electrical signals. The lenti or adeno associated (AAV) viral vectors containing the gene construct injection at the region of desired cells population and Cre-Lox P recombination system are popular methods of transfection. For the viral approach there are two possible viral approaches: selecting a virus that infects specific a cell type or incorporating a cell-type specific promoter sequence in gene construct to restrict the opsin protein expression in a particular cell type. Optogenetics is always coupled with microscopy such as two-photon or multi-photon, electrophysiology, or behavioral studies. In neuroscience, the specific types of neurons are transfected with an opsin gene containing vector and the light stimulating opto fibres are placed near the site of neurons to be triggered or observed. After triggering a specific set of neurons, the behavior of the animal is assessed. Also, the activated network of neurons at the time of particular behavior can be checked by electrophysiological studies and attached computational tools (Miesenböck 2005; Deisseroth 2011).

Many applications are emerging with research on different aspects of the method. The neural circuits involved in the regulation of behavioral states have been studied in the nematode, fruit fly, song bird, zebra fish and non-human primates. Deciphering the neuronal circuits underpinning anxiety, fear and other psychiatric disorders would help to diagnose a disease and the cause of particular behavior and will unveil new therapeutics to treat them (McDevitt 2014). Botond Roska, and Jose-Alain Sahel, a neuroscientist and ophthalmologist respectively, are working on establishing optogenetic method to restore sight in blind persons. Apart from the application of optogenetics in neuronal studies, Michael Lin's lab at Stanford University is creating light-controlled proteins to study nucleotide exchange and protease activities.

References

Boyden, Edward S. 2011. "A History of Optogenetics: The Development of Tools for Controlling Brain Circuits with Light." F1000 Biology Reports. 3:11. http://f1000.com/prime/reports/b/3/11. Accessed on May 10, 2016.

Deisseroth, Karl. 2011. "Optogenetics." *Nature Methods*. 8: 26–29.

Deisseroth, Karl. 2015. "Optogenetics: 10 Years of Microbial Opsins in Neuroscience." *Nature Neuroscience*. 18: 1213–1225.

Gorostiza, Pau, and Ehud Y. Isacoff. 2008. "Optical Switches for Remote and Noninvasive Control of Cell Signaling." *Science*. 322(5900): 395–399.

McDevitt, Ross A., Sean J. Reed, and Jonathan P. Britt. 2014. "Optogenetics in Preclinical Neuroscience and Psychiatry Research: Recent Insights and Potential Applications." *Neuropsychiatric Disease and Treatment*. 10: 1369–1379.

Miesenböck, Gero, and Ioannis G. Kevrekidis. 2005. "Optical Imaging and Control of Genetically Designated Neurons in Functioning Circuits." *Annual Review of Neuroscience.* 28: 533–63.

Zhang, Feng, et al. 2011. "The Microbial Opsin Family of Optogenetic Tools." *Cell.* 147(7): 1446–1457.

Arpita Dave is pursuing her PhD in neurobiology from the M. S. University of Baroda, India. She is an avid reader of emerging technologies and electrophysiological methods for understanding how the brain functions.

Africans Using Autosomal DNA Testing to Find Distant Family Members
LaKisha David

Between the 15th and 19th centuries, 11 to 13 million Africans were forced from West, West Central, and Southeast Africa and sailed to Americas as captives, headed toward a life of chattel slavery (Drake 1975; Jackson and Borgelin 2010). These Africans were dispersed to various countries in North America, the Caribbean, and South America. It is reasonable to assume that every person chained as cargo on those transatlantic voyages desired to return to their families in Africa.

For most African Americans, most of whom are descendants of Africans enslaved in the United States, ethnic identity begins with slavery in this country and awareness of having ancestors from Africa (Sellers et al. 1998). Ethnic identity refers to a "sense of self as a member of an ethnic group . . . that claim[s] a common ancestry and share one or more of the following elements: culture, phenotype, religion, language, kinship, or place of origin" (Phinney 2003, 63). Many African Americans seek to build their sense of self through identification with Africa. For this segment of the African American population, the goal is to build their ethnic identity through discovering their lost African ancestry. DNA testing often aids in this process (Nelson 2008; Winston and Kittles 2005). There are a growing number of African Americans who conduct DNA testing to determine their ethnic group or country of origin in Africa (Jackson and Borgelin 2010).

DNA testing to discover ancestry usually comes in the form of ethnic lineage tests such as YDNA or mitochondrial DNA (mtDNA) testing (Nelson 2008). The mtDNA test examines the maternal lineage, the male or female tester's mother's lineage going back for thousands of years. Similarly, YDNA test examines the paternal lineage, the male tester's father's lineage going back for thousands of years. One commercial company that specializes in using mtDNA and YDNA testing to help people of African descent discover African ethnic groups related to their ancestors is African Ancestry. Despite the debate related to the accuracy of using the YDNA and mtDNA tests to identify ethnic origins, African Ancestry enjoys increasing patronage.

Although DNA reference samples for YDNA and mtDNA tests were typically based on ethnic groups tens of thousands of years ago (Jackson and Borgelin 2010), African Ancestry reports producing accurate results because they focus solely on African ancestry. They reportedly have developed the largest African DNA reference database by methods which include having tested individuals throughout Africa (African Ancestry 2015). With improvements in technology and methods, ancestry DNA companies are now able to compare customer DNA to samples 500 to a few thousand years ago instead of the tens of thousands of years ago of earlier test results (Jackson and Borgelin 2010). This is particularly significant for the African Americans who face enormous difficulty in determining direct lineages to specific ethnic groups by traditional means in the United States. Scholars such as Alondra Nelson conclude that testing based on mtDNA and YDNA provides an incomplete knowledge for discovering ethnic identity.

The autosomal DNA (atDNA) test eliminates most of this educated guesswork within certain thresholds by allowing companies to compare their customers' DNA to each other to identify degrees of relatedness. By this means, people of African descent who are related through unknown enslaved African ancestors can reconnect and attempt to piece together their shared history. For example, if an Irish American did an atDNA test, this individual's DNA would be matched with all other customers' DNA and a list of possible relatives would be displayed. A person with 0.195 percent of DNA in common (or about 13.28 centimorgans of DNA in common) would appear as a fourth cousin in the DNA relatives results. If a first-generation African did an atDNA, he or she would be matched with their African American distant cousins within the customer database. This is significant because families, not just vague notions of ethnic groups, who were torn apart through the transatlantic slave trade can now be reunited.

Through African/African Diaspora distant cousin reunification, there are many opportunities for information sharing and making family. Autosomal DNA testing has the potential to redefine the way Africans (continental Africans, immigrant Africans, people of African descent) claim family and community. Some may critique this experience because biological kinship is being essentialized. However, people of African descent have a tendency to base kinship on non-biological as well as biological connections. If this trend holds, adding new information about the biological kinship between African immigrants in the United States and African Americans could lead to greater integration and stronger communities among the African diaspora in the United States. Autosomal DNA testing is restoring families who were torn apart through the transatlantic slave trade and creating a greater sense of self for African American testers.

References

African Ancestry. 2015. "African Ancestry—Trace Your DNA. Find Your Roots. Today." http://www.africanancestry.com/home/. Accessed on May 15, 2016.

Drake, St Clair. 1975. "Black Diaspora in Pan-African Perspective." *Black Scholar* 7(1): 2–13.

Jackson, Fatimah L. C., and Latifa F. J. Borgelin. 2010. "How Genetics Can Provide Detail to the Transatlantic African Diaspora." In *The African Diaspora and the Disciplines*, edited by Tejumola Olaniyan and James H. Sweet. Bloomington: Indiana University Press.

Nelson, Alondra. 2008. "Bio Science Genetic Genealogy Testing and the Pursuit of African Ancestry." *Social Studies of Science* 38(5): 759–783.

Phinney, Jean S. 2003. "Ethnic Identity and Acculturation." In *Acculturation: Advances in Theory, Measurement, and Applied Research*, edited by K. M. Chun, P. Balls Organista, and G Marin. Washington, DC: American Psychological Association. http://psycnet.apa.org/index.cfm?fa=main.doiLanding&uid=2002-18425-006. Accessed on May 15, 2016.

Sellers, R. M., et al. (1998). "Multidimensional Model of Racial Identity: A Reconceptualization of African American Racial Identity." *Personality and Social Psychology Review*, 2(1), 18–39.

Winston, Cynthia E., and Rick A. Kittles. 2005. "Psychological and Ethical Issues Related to Identity and Inferring Ancestry of African Americans." In *Biological Anthropology and Ethics: From Repatriation to Genetic Identity*, edited by Trudy R. Turner. Albany: State University of New York Press.

LaKisha David is a PhD student in the Human Development and Family Studies (HDFS) Department of the University of Illinois at Urbana-Champaign (UIUC) studying the social and psychological impact on African family reunification. She is also the director of the African Kinship Reunion (TAKiR), whose mission is to reunite African families separated by the transatlantic slave trade.

Mandatory GMO Labeling Is the "Right to Be Deceived"
Jon Entine

Labeling of foods with genetically modified ingredients is now a reality. When it became clear that legal challenges would not be resolved in time to prevent the implementation of the Vermont law, numerous companies began voluntary labeling. Campbell, ConAgra, Kellogg, Mars, Del Monte and other companies either began labeling some of their products or will soon do so.

These companies have voluntarily put some variation of "contains genetically engineered ingredients" on the back panel near the mandated ingredients box, which is similar to the way Europe labels. But many biotechnology critics would rather see "GMO" mandated—an ideological acronym loathed by scientists but hatched years ago by activists to create the scary image of foreign "organisms" crawling in our food. They also want to put the label on the package front. That prominence would serve as kind of skull and crossbones, raising questions among consumers about health and safety.

Yet, every prominent science organization in the world—more than 270 to date—has publicly stated there are no unique health hazards associated with genetically engineered foods. In May 2016, an expert committee convened by the National Academies issued its long-awaited assessment of GE foods, concluding there are "no differences that would implicate a higher risk to human health and safety than from eating their non-GE counterparts" and that it "does not believe that mandatory labeling of foods with GE content is justified to protect public health."

What are the arguments mustered for and against labeling? Critics say it is about one thing: the "right to know." But is that really their motive? Andrew Kimbrell, head of the Center for Food Safety, which acts as a legal SWAT team for anti-GMO campaigners, wrote a book titled *Your Right to Know*. But he's revealed his real motivation elsewhere, writing, "We are going to force them to label this food. If we have it labeled, then we can organize people not to buy it."

To pro-labeling advocates, GMOs are a symbol of "industrial agriculture." A GMO label would convey "Big Ag" to the consumer with all the negative connotations that foodies associate with that term. But that is not accurate. More than half of all GE crops are grown in the developing world by small farmers. But the "industrial farming" myth is so powerful and so widely embraced that it powers the labeling movement.

Opponents of mandatory labeling say nothing, scientifically speaking, can be gained from a GE label. Scientists, including at the FDA, contend there is no "material difference" between GE and non-GE foods. Most importantly, nutrition content is the same or lower in non-GMO foods, which often strip out many vitamins, such as B12, riboflavin, and vitamin C, because they are made with genetically engineered enzymes.

Although the non-GMO label does not allow its label to be used on products with GE vitamins, Vermont law (and those proposed in other states) set different rules. They do not require labeling of iconic Vermont cheeses (and breads and baked goods and beer and juices), which are made with GE enzymes. Nor will labels be required on soft drinks, which are made with fructose derived from GE sugar. Nor beef, pork, and chicken, although almost all feed animals throughout the developed world (yes, even in Europe), are raised on GE grains.

Why? Because Vermont legislators apparently believed that labeling sodas and burgers or gourmet cheeses might scare consumers, although labeling other GE foods would not have the same effect. And, because of science there are no traces of the GE ingredients in the final product made with GE microbes.

But if "no traces" is the standard used in Vermont, then why is sugar made from GE sugar beets labeled? Sugar contains no DNA at all and cannot be distinguished in a laboratory from non-GMO sugar made from sugarcane. But anti-GMO activists convinced legislators that some GE crops should be labeled based on the seed that was used to grow it, rather than the final product, which like all GE foods, is "GMO free." As a result, companies that want to avoid putting a GE label on products with fructose cannot use GE sugar, which is grown on American farms with almost no use of insecticides. Instead, they will have to import their sugar at considerable expense and ecological waste from Latin America or Asia, where insecticides are heavily used and ecological standards are low.

Does not the consumer have a "right to know" about these science and sustainability contradictions?

Most geneticists and legal experts oppose mandatory GE labels. Federal food labels have a defined purpose: providing nutritional and safety information to consumers. This includes things that can be measured—calories, fats, vitamins, mineral content, potential allergens—and methods of preparation that can impact these measureable ingredients. (FDA calls this "material composition.")

A second role of government in labeling is to ensure truth in advertising. Labels identifying products as organic, kosher or halal, fall into this category. They were developed by industry to address consumer preferences but are overseen by government agencies to ensure truthfulness and accuracy.

GMO ingredients fit neither category. GMO is not an ingredient but a process to develop hybrid seeds, which have been used and patented since the 1930s, to improve taste, nutrition or disease resistance. So, "calling out" foods from crops whose seeds are genetically engineered tells the consumer . . . nothing of substance. The National Academies reinforced this point in its landmark analysis, concluding that regulations should focus on the final product of both conventional and GE breeding and novel characteristics not the process.

It is important to note that consumers are not informed of other kinds of far more drastic breeding modifications, such as mutagenesis. That is a process, developed in the 1930s and carried out in laboratories, that randomly blows out thousands of chromosomes of a seed using radiation or chemicals. There are more than 3,000 foods and plants that have been made this way, including Ruby and Rio red grapefruits and Italian pasta-grade durum, both of which are sold as organic, with no labeling as to the process. Yet, food from GM seeds, which tweak one or two genes, is labeled.

Mandatory labeling would have other consequences as well. As it would set "contamination" levels, it would require every part of the production chain-grower, transporter, and processer—to set up segregated dual systems for GM and non GM crops. That is expensive. And it would open up the legal floodgates to tort actions from anti-GMO activists, like Andrew Kimbrell.

The main consequence would be that risk-averse companies would curtail producing GE foods—which is what the "right to know" forces really want after all—reducing consumer choice.

Jon Entine is founder and director of the Genetic Literacy Project, an independent nonprofit that reports on the intersection of genetics, journalism, and policy, and Senior Fellow at the University of California-Davis Institute for Food and Agricultural Literacy. He is a longtime television and print journalist, winning 19 major awards, author of 7 books, and a frequent commentator on genetics and biotechnology.

A Brave New World for CRISPR/Cas9: Scientific Limitations and Ethical Considerations
Rachele Hendricks-Sturrup

Introduction
Novel gene-editing technologies, such as clustered regularly interspaced short palindromic repeats (CRISPR)/Cas9, hold a global spotlight in the scientific community. For decades, scientists have explored several mechanisms in which genomic-level editing technologies can be used to produce favorable changes across diverse species and cellular types,

such as embryonic stem cells of the human species. Thousands of genes within the human genome are linked to a number of human ailments, such as cancer, and through modification of those genes, CRISPR/Cas9 could potentially interrupt the pathological sequence of events caused by such genetically derived ailments. Further and more compelling, CRISPR/Cas9 technology can be used to modify genomes of embryonic stem cells, which could cause an establishment of heritable changes across subsequent generations produced by that particular germline.

CRISPR/Cas9 is lauded globally for its relative efficiency and affordability. However it is important to understand and address its scientific limitations and ethical considerations. These include but are not limited to CRISPR/Cas9's ability to create and propagate unnatural selection, and the risk of inappropriate use of CRISPR/Cas9 technology. This article provides a general overview of the CRISPR/Cas9 technology, discusses some of its major scientific limitations and ethical considerations, and lastly provides recommendations to address those limitations and considerations moving forward.

CRISPR/Cas9: A Brief Scientific Overview

CRISPR/Cas9 is an acquired immune system variant found naturally in several genomes of bacteria and archaea against invading viruses and phages. CRISPR, a DNA locus enclosing a series of conserved and repeated sequences interspaced by non-repetitive and distinct spacers, conjoins with various Cas nucleases, such as Cas9 or Cas4, to produce transcriptional templates of crRNA that are used downstream to target and degrade invading virus and phage DNA (Lemak et al. 2014). To date, over 40 different Cas nuclease families have been reported as constituents of prokaryotic CRISPR/Cas systems (Haft et al. 2005). The crystal structure of *Streptococcus pyogenes* Cas9 in complex as an asymmetric unit with guide RNA and target DNA was recently solved by Nishimasu et al. through X-ray diffraction at 2.5 Å (2014). CRISPR/Cas9-derived technology harnesses the power and specificity held by cellular machinery within bacteria and archaea, and uses this machinery to selectively modify genomes of either prokaryotic or eukaryotic cells, as desired (Ran et al. 2013; Shalem et al. 2014; Cong et al. 2013).

Scientific Limitations and Ethical Considerations

Genomic manipulation is almost always accompanied by ethical concerns and debate over eugenics, as well as scientific limitations that may expose the risk of unpredictable outcomes that could manifest at various organization levels of an organism and beyond. Scientific limitations to CRISPR/Cas9 technology include unknown evolutionary effects caused by interference with natural selection, the risk of incomplete genomic-level editing that could occur during the CRISPR/Cas9 process, and the risk of Cas9 DNA degradation of genes homologous, but not equivalent, to target genes. Those interested in pursuing practical translation of CRISPR/Cas9 must understand, inform, and directly address the risk of improper and irresponsible use, such as modifying genomes to vainly enhance physical appearances (or cosmetic uses), loss of genetic diversity due to eugenics, and the inability to preemptively obtain informed consent from human offspring bearing CRISPR/Cas9 modified genomes. Ethical concerns also span across the ongoing and global debate against embryonic stem cell use in biomedical research, which includes CRISPR/Cas9 technology research (Baltimore 2015; Otieno 2015; Rodriguez 2016).

Recommendations

In April 2015, the U.S. National Institutes of Health's (NIH) Director, Francis Collins, made a public statement against CRISPR/Cas9 research in human embryos, and shared the fact that many legislative and regulatory prohibitions exist against such research (Collins 2015). Following this statement in December 2015, the National Academy of Sciences convened an international group of scientists in Washington D.C. to discuss the appropriateness of CRISPR/Cas9 research on human genes. During this meeting, David Baltimore, former president of the California Institute of Technology, stated, "I very much hope that people will understand that this is a major step forward, but like any new technology, it has to be approached with care and thoughtfulness" (Baltimore et al. 2015).

In consideration of the previous comments and discussion, it is recommended that scientists and bioethicists continue to hold international moratoriums to discuss responsible CRISPR/Cas9 research and use, and, under the auspices of transparency, allow public access those moratoriums. Moreover, scientists should consider and discuss alternative, and less controversial, approaches to CRISPR/Cas9 stem cell research, which may include investigating CRISPR/Cas9 genome editing in non-viable embryonic stem cells and/or somatic stem cells carrying disease-linked genes. Despite its limitations, CRISPR/Cas9's unparalleled fidelity at the genomic level, relative affordability, and overall simplicity illuminate CRISPR/Cas9 as a spotlight target in our brave new world that seeks to embrace biotechnology innovation and translation.

References

Baltimore, David, et al. 2015. "A Prudent Path Forward for Genomic Engineering and Germline Gene Modification." *Science.* 348(6230): 36–38.

Collins, Francis S. 2015. "Statement on NIH Funding of Research Using Gene-Editing Technologies in Human Embryos." U.S. National Institutes of Health. April 29, 2015. https://www.nih.gov/about-nih/who-we-are/nih-director/statements/statement-nih-funding-research-using-gene-editing-technologies-human-embryos. Accessed May 1, 2016.

Cong, Le, et al. 2013. "Multiplex Genome Engineering using CRISPR/Cas Systems." *Science*. 339(6121): 819–823.

Haft, Daniel H., et al. 2005. "A Guild of 45 CRISPR-Associated (Cas) Protein Families and Multiple CRISPR/Cas Subtypes Exist in Prokaryotic Genomes." *PLoS Computational Biology*. 1(6): e60.

Lemak, Sofia, et al. 2014. "The CRISPR-Associated Cas4 Protein Pcal_0546 from *Pyrobaculum calidifontis* Contains a [2Fe-2S] Cluster: Crystal Structure and Nuclease Activity." *Nucleic Acids Research*. 42(17): 11144–11155.

Nishimasu, Hiroshi, et al. 2014. "Crystal Structure of Cas9 in Complex with Guide RNA and Target DNA." *Cell*. 156(5): 935–949.

Otieno, M. O. 2015. "CRISPR-Cas9 Human Genome Editing: Challenges, Ethical Concerns and Implications." *Journal of Clinical Research & Bioethics*. 6(6). doi: 10.4172/2155-9627.1000253.

Ran, F. Ann, et al. 2013. "Genome Engineering Using the CRISPR-Cas9 System." *Nature Protocols*. 8(11): 2281–2308.

Rodriguez, E. 2016. "Ethical Issues in Genome Editing Using Crispr/Cas9 System." *Journal of Clinical Research & Bioethics*. 7(2). doi:10.4172/2155-9627.1000266.

Shalem, Ophir, et al. 2014. "Genome-Scale CRISPR-Cas9 Knockout Screening in Human Cells." *Science*. 343(6166): 84–87.

Wade, Nicholas. 2015. "Scientists Seek Moratorium on Edits to Human Genome That Could Be Inherited" *New York Times*. December 3, 2015. http://www.nytimes.com/2015/12/04/science/crispr-cas9-human-genome-editing-moratorium.html?_r=0. Accessed May 01, 2016.

Rachele Hendricks-Sturrup is a biomedical scientist and active member of the Association for Women in Science, American Medical

Writers Association, and National Association of Science Writers. She is currently pursuing a Doctor of Health Science degree at Nova Southeastern University. Her doctoral study focus involves exploring and describing how various forms of biotechnology can be used within the scope of precision medicine and value-based care.

Genome Editing Opens Brave New World
Nerissa Hoglen

From *Brave New World* to *Gattaca*, the repercussions of gaining genetic control over individuals' traits are a preoccupation of science fiction. The recently developed CRISPR/Cas9 genome-editing technique is now making those concerns a preoccupation of science-nonfiction necessary.

Genome editing refers to the process of making specific changes to the DNA of an organism in order to change a trait. An organism's DNA provides information that specifies the structure of proteins, the function of which determines the traits of the organism. The CRISPR/Cas9 technique is a rapid way to make changes directly to DNA—and as a result, proteins—in a variety of organisms ranging from plants to primates.

The CRISPR system makes use of two components: a piece of DNA provided by the experimenter that indicates which DNA in the organism to target, and Cas9, a protein that catalyzes the change to the target region of organismal DNA. Using this targeting/editing system, a scientist can add and remove small regions of DNA. By changing the DNA, scientists can make changes to the proteins that will be made from the DNA, ultimately altering traits in the organism. Like many tools biologists have engineered, this is a system that bacteria use as a defense against viral invaders.

Although genome editing was possible before the CRISPR system, CRISPR has made the process much faster and easier, and it is proliferating in research laboratories. Using older methods, it is tricky to design the desired edit, and it is necessary to breed many generations of transgenic organisms before the edit is expressed robustly. Genome editing allows scientists do things like engineering mice to have genetic mutations linked to disease in humans so they can research causes and therapies of that disease without testing on patients. Increasing the speed of this process means science can help more patients. CRISPR is already helping to make strides in science and medicine, offering huge benefits to society.

The increased efficiency of the CRISPR method also raises the possibility of using genome editing in humans because it can be applied directly to somatic cells, in contrast to older methods that target germline cells. Before CRISPR, scientists edited the DNA of reproductive cells in order to make genetic changes. These edits were not expressed in the original animal but would appear in subsequent generations. However, CRISPR allows scientists to directly target non-reproductive cells, which means CRISPR could be used to treat human conditions. For example, it may be possible to use the CRISPR/Cas system to delete rogue DNA that causes malignant cancer cells to replicate inappropriately. This could revolutionize cancer treatments, bringing science into the realm of amazing technology anticipated in science fiction and helping countless people.

However, the potential to edit human genomes is a large responsibility and raises ethical concerns about how to use this technology. There are three major concerns scientists and the public need to address: the heritability of edits, the complexity of targeting edits, and unforeseen consequences of using CRISPR.

Using CRISPR to edit human genomes raises the possibility of making mutations that can be passed down from one generation to the next. Unlike the proposed use for editing the DNA in cancer cells, CRISPR could also be used to make germline edits that would be passed on to children, grandchildren, and so on. It is controversial whether anyone has the right to make such far-reaching genetic changes, especially for cosmetic (as opposed to health-related) reasons. This possibility is different from using CRISPR to stop cancerous cells because those edits would be targeted solely to the cancer cells and would not be passed down to future generations.

A further complication of using CRISPR in humans is the complexity of the genetic basis of the traits being altered and the limited scope of the edits CRISPR can make. Until our understanding of genetics has caught up with our ability to edit genes, we will not be able to make well-designed edits. Because we do not know the effects of a given genetic edit or the combined effects of changing genes that work in concert, an edit could have disastrous unintended consequences.

Even a well-designed edit could have as-yet unidentified long-term health effects. Many questions about safety remain unanswered. Scientists cannot predict all possible downstream effects of modifying a gene in the genome, meaning that extensive testing in laboratories is necessary before trying to apply the technique to humans. A small number of laboratories have used CRISPR to edit the genomes of primates or non-viable human embryos, but many prominent scientists have called for a moratorium on experiments in human tissue or primates until more is known (Baltimore et al. 2015; Doudna 2015).

The potential for the CRISPR/Cas9 technique to revolutionize biological research and medicine is thrilling: it has already increased the pace of biological experiments and offers the possibility of precisely targeted interventions for cancer and genetic disorders. These benefits could help us achieve some of the utopian goals of science fiction. However, scientists and the public also need to work together to use CRISPR technology ethically and avoid realizing any of science fiction's darker predictions.

References

Baltimore, David, et al. 2015. "A Prudent Path Forward for Genomic Engineering and Germline Gene Modification." *Science.* 348(6230): 36–38. doi:10.1126/science.aab1028.

Doudna, Jennifer. 2015. "Genome-Editing Revolution: My Whirlwind Year with CRISPR." *Nature.* 528(7583): 469–71. doi:10.1038/528469a.

Nerissa Hoglen is a graduate student studying auditory processing of social and other natural sounds at the University of California, San Francisco.

Finding a Face in the DNA
Clara MacCarald

In 2015, a laboratory created the first computer-generated image of a murder suspect using only crime-scene DNA (Wolinsky 2015, 1). The technique, forensic DNA phenotyping, is showing promise not only in the field of criminal forensics, but also in archeology. Researchers caution the technology is still new, while privacy advocates worry about potential for misuse as researchers dig into the real-world expression of our genetic code (Wolinsky 2015, 1; Pollack 2015).

Traditional DNA identification requires one known sample. The technique compares segments of the genome where base pair sequences repeat themselves. The number of repeats varies among people and is inherited from both parents. These repeated base pairs have no known function and therefore give no independent information about the person who left the DNA (Greely et al. 2006, 249–250).

Forensic DNA phenotyping is different. It aims to read the genetic code itself. Laboratories are exploring how variation in peoples' genotype relates to their phenotype, or expressed traits. Determining visible traits from a scrap of body tissue might help analyze a DNA sample when investigators lack a DNA match or eyewitnesses (Albert and Wright 2015, 1–2).

Phenotyping looks for places in the genome where individuals vary by one base. This is called single nucleotide polymorphism (SNP). The goal is to find SNPs that are correlated with observable variations in peoples' phenotypes, especially in their faces. Genetics determine many of our recognizable facial features. Family members typically resemble each other, and identical twins may be indistinguishable (Claes et al. 2015, 208).

People within ethnic groups also resemble each other. Researchers have used SNPs to predict ancestry with 98.6 percent accuracy in a test population (Albert and Wright 2015, 4). Sex and ancestry are the strongest predictors researchers have found so far for a person's facial appearance. Researchers have been searching for SNPs that can fine-tune such features as nose or mouth width (Claes et al. 2015, 210–211).

One fertile field of inquiry has been in pigments. For example, by focusing on SNPs related to eye color, researchers were able to create a test called IrisPlex. The test is 95 percent accurate in predicting blue or brown eyes. Intermediate colors are harder. The same researchers also created HIrisPlex, which predicts hair color with accuracies of 80 percent to 93 percent depending on the color (Wolinsky 2015, 2).

Researchers used HIrisPlex to help picture one medieval individual. Existing portraits of King Richard were all painted after his death in 1485 and show different facial colorings. In 2012, the suspected remains of Richard III were found. Researchers confirmed the identity using skeletal features and mitochondrial DNA, which matched a living relative of the historic king (King et al. 2014). Phenotyping predicted that Richard III probably had blue eyes and blond to brown hair. This matched a specific portrait in London's Societies of Antiquaries (Cookson 2015).

Genetics are only one component of our faces, while diet, disease, sun exposure, and accidents leave traces on our appearance that are not recorded in our genes (Claes et al. 2015, 209; Albert and Wright 2015, 3). Age affects both appearance and DNA. Some attempts to predict age have not worked out, such as by measuring telomeres, which are repetitive sequences at the end of chromosomes that shorten over time. Researchers are currently working on estimating age based on a mix of DNA changes (Wolinsky 2015, 3).

Small companies that perform phenotyping have started offering their services to law enforcement agencies (Pollack 2015). Sex and ancestry may still be the most helpful information to law enforcement (Wolinsky 2015, 1). In an early application, phenotyping played a pivotal part in redirecting an investigation.

In 2003, Louisiana police were searching for a serial killer. Research suggests most serial killers are Caucasian males, and witnesses in this case seemed to confirm this assumption. But a genetics testing company determined that DNA from the crime scene had 85 percent sub-Saharan African ancestry. The new focus led to the arrest of a black man who was eventually convicted (Pollack 2015).

Some researchers dismiss the first wanted poster created solely from DNA as a generic black male's face. In this case, the analysis of DNA from the location of the double homicide gave an ancestry of 92 percent West African and 8 percent European (Wolinsky 2015, 1). Some people worry this reliance on genetic ancestry could worsen racial profiling (Pollack 2015). Privacy advocates are also concerned as researchers uncover genetic susceptibilities to disease and, perhaps one day, even addictions (Allocca 2015).

Investigators and laboratories counter that the characteristics they focus on are visible to anyone who looks at a suspect and, therefore, public knowledge (Allocca 2015). Others point to the exclusive use of phenotyping as an investigative tool. Trials rely on traditional DNA identification (Wolinsky 2015, 5). Despite the limitations and risks, phenotyping is not going away. Researchers continue to search for more traits to predict (Wolinksy 2015, 3–4). And law enforcement continues to turn to phenotyping, hoping to warm up cold trails (Pollack 2015).

References

Albert, A. Midori and Charissa L. Wright. 2015. "DNA Prediction in Forensic Anthropology and the Identity Sciences." *Global Journal of Anthropology Research*. 2: 1–6. http://www.cosmosscholars.com/phms/index.php/gjar/article/viewFile/390/244. Accessed on March 30, 2016.

Allocca, Sean. 2015. "First DNA-Phenotyped Image of 'Person of Interest' in Double Homicide." Forensic Magazine. http://www.forensicmag.com/article/2015/01/first-dna-phenotyped-image-person-interest-double-homicide. Accessed on August 26, 2016.

Claes, Peter, Harold Hill, and Mark D. Shriver. 2014. "Toward DNA-Based Facial Composites: Preliminary Results and Validation. *Forensic Science International: Genetics*. 13: 208–216. https://www.researchgate.net/profile/Peter_Claes/publication/265387983_Toward_DNA-based_facial_composites_Preliminary_results_and_validation/links/5410084c0cf2d8daaad0c1dd.pdf. Accessed on April 1, 2016.

Cookson, Clilve. 2015. "DNA: The Next Frontier in Forensics." FT Magazine. http://www.ft.com/cms/s/2/012b2b9c-a742-11e4-8a71-00144feab7de.html. Accessed on August 26, 2016.

Greely, Henry T., et al. 2006. "Family Ties: The Use of DNA Offender Databases to Catch Offenders' Kin." *The Journal of Law, Medicine, and Ethics*. 32: 248–262. http://law.stanford.edu/wp-content/uploads/sites/default/files/event/264460/media/slspublic/Greely.pdf. Accessed on April 10, 2016.

King, Turi E., et al. 2014. "Identification of the Remains of King Richard III." *Nature Communications*. 5: 5631. http://www.nature.com/ncomms/2014/141202/ncomms6631/full/ncomms6631.html. Accessed on April 10, 2016.

Pollack, Andrew. 2015. "Building a Face, and a Case, on DNA." *New York Times*, February 23. http://www.nytimes.com/2015/02/24/science/building-face-and-a-case-on-dna.html?_r=1. Accessed on March 30, 2016.

Wolinsky, Howard. 2015. "CSI on Steroids." *EMBO Reports* 16: 782–786. http://www.iupui.edu/~walshlab/Media/Offline%20Webpages/CSI-EMBO%20reports.pdf. Accessed on April 10, 2016.

Clara MacCarald is a freelance writer with an MS in biology who lives in Central New York. She is currently writing several educational books for children.

CRISPR Controversies
Deirdre Manion-Fischer

CRISPR-Cas9 is faster, cheaper, and more specific than previous gene-editing strategies. The technology has not won a Nobel Prize yet, but it almost undoubtedly will, since it has profoundly changed the field since it appeared on the scene about three years ago.

An ongoing dispute over patents for this technology involves two prestigious institutions: The Broad Institute, associated with MIT and Harvard, and the University of California-Berkeley. Research groups at each institution have made significant contributions to the invention of CRISPR-Cas9. Jennifer Doudna, of Berkeley, showed that CRISPR works in purified bacterial DNA and published first (Jinek et al. 2012), while Feng Zhang of the Broad Institute showed that it was effective in human cells (Hsu et al. 2013). The patent was awarded to Broad in April 2014, and the University of California disputed the decision (Begley 2016).

No matter who wins, researchers will be able to use the technology without having to pay licensing fees. However, any company that wants to develop CRISPR for commercial use will have to pay them. Millions of dollars are at stake for the respective institutions. So far, three companies have licensed the technology to develop its therapeutic potential.

Scientists and philosophers have discussed the ethics of using gene-editing technology even before such technology existed. A famous conference was held in Asilomar, California, in 1975, at which scientists, lawyers, and other interested individuals established guidelines for using recombinant DNA safely. With CRISPR, these concepts have been revisited many times.

Eighteen scientists involved in DNA research, including Jennifer Doudna, published a letter in *Science* (Baltimore et al. 2015) in which they recommended that steps be taken to "strongly discourage . . . any attempt at genome modification for clinical application in humans." Their reasoning and main concerns are based on the fact that we have limited knowledge of human genetics that could result in unintended consequences from previously unknown interactions with other diseases and the environment.

Michael Specter described such an example in his November 2015 *New Yorker* article: a single mutation, the Dlta32 mutation in cell membrane protein CCR5 prevents HIV (that causes AIDS) from entering the cell. However the mutation also increases susceptibility to the West Nile Virus (Specter 2015).

Baltimore and co-authors of the letter stated that more research is needed into unintended consequences. They said also that there is an "urgent need for open discussion of merits and risks of genome modification by a broad cohort of scientists, clinicians, social scientists, [and] the general public." They recommended convening such a globally representative group. They believe that the FDA regulations governing the Investigational New Drug (IND) Exemption, which is required in the United States to administer any drug to humans, are insufficient for the application of CRISPR-Cas9 genome-editing techniques.

There has been some research into the effect of using CRISPR-Cas9 to edit the DNA of human embryos. When the first such experiment was carried out in China, about a year ago, there was an uproar in the press, which has since died down. The scientists attempted to repair the gene responsible for beta thalassemia, a blood disorder, in 86 human embryos (Liang et al. 2015). The embryos they used are produced but cannot be used in the course of in vitro fertilization. In the United States, federal funding for such research is not allowed, but it is approved in Sweden and the United Kingdom. Now there is more broad scientific support for research using embryos, for its potential to improve CRISPR technology. The Chinese experiment produced a lot of genetic errors, so there is a long way to go.

With CRISPR, it is now possible to make what is called a "gene drive" that can copy itself into the corresponding allele of offspring, so it has close to 100 percent likelihood of inheritance, as opposed to 50 percent as normal. This technique has now been used to make a genetically engineered population of mosquitoes resistant to a pathogen responsible for malaria (Gantz et al. 2015). One concern is environmental consequences, such as the destruction of a potentially important population of insects that would have a negative impact on the animals that use them as a food source.

One final promising use of CRISPR is to help with the current shortage of human organs available for transplantation by implementing the use of pig organs. One problem with this is that pig genomes contain viral DNA that could infect humans. By integrating the cutting enzyme in CRISPR into the pig genome and then guiding it to 62 viral genes with RNA, a research team was able to inactivate all of them, a step on the road to save hundreds of patients' lives per year (Yang et al. 2015).

References

Baltimore, David, et al. 2015. "A Prudent Path Forward for Genomic Engineering and Germline Gene Modification." *Science*. 348(6230): 36–38.

Begley, Sharon, "In the CRISPR Patent Fight, the Broad Institute Gains Edge in Early Rulings," *STAT News*, March 18, 2016. https://www.statnews.com/2016/03/18/crispr-patent-dispute/. Accessed May 15, 2016.

Gantz, Valentino M., et al. 2015. "Highly Efficient Cas9-Mediated Gene Drive for Population Modification of the Malaria Vector Mosquito *Anopheles stephensi.*" *Proceedings of the National Academy of Sciences.* 112(49): E6736–E6743.

Hsu, Patrick D., et al. 2013. "DNA Targeting Specificity of RNA-Guided Cas9 Nucleases." *Nature Biotechnology.* 31(9): 827–832.

Jinek, Martin, et al. 2012. "A Programmable Dual-RNA— Guided DNA Endonuclease in Adaptive Bacterial Immunity." *Science.* 337(6096): 816–822.

Liang, Puping, et al. 2015. "CRISPR/Cas9-Mediated Gene Editing in Human Tripronuclear Zygotes." *Protein & Cell.* 6(5): 363–372.

Specter, Michael. 2015. "The Gene Hackers." *New Yorker.* November 16, 2015.

Yang, Luhan, et al. 2015. "Genome-Wide Inactivation of Porcine Endogenous Retroviruses (PERVs)." *Science.* 350(6264): 1101–1104.

Deirdre Manion-Fischer earned an MS in chemistry in 2012 under the direction of Edgar Arriaga at the University of Minnesota-Twin Cities, and a BS in chemistry from Kent State University in 2010. She is a freelance science writer living in Minneapolis, Minnesota, with her cat, Twist.

Modified DNA for Targeted Therapeutics
Manish Muhuri

The DNA molecule has been one of the most important sources of information for understanding the fundamental basis of human life and evolution. Elucidation of the double helical DNA structure and unraveling of the human genome sequence are two landmarks of 20th-century genetics. The Human Genome Project (HGP) provided a major impetus in identifying the underlying causes of many genetic diseases. However, greater efforts are needed in order to translate this knowledge to target the genes implicated in genetic diseases for therapeutic purposes (Venter 2001). More than 10,000 human diseases are estimated to have genetic abnormalities as the underlying cause. The HGP helped scientists determine genetic markers that are responsible for individual drug response and their potential side effects.

In clinical medicine, development of antibodies has revolutionized therapeutics and improved the quality of life of patients. Unfortunately, these antibodies sporadically arouse unwanted immune responses to themselves. Moreover, quality control in mass production can also be problematic. Using genomic data, however, DNA-based drugs can be developed for individualized medicine. Moreover, the HGP has heralded a new era for treating diseases, the development of DNA-based drugs for gene replacement, gene therapy, or gene ablation, a process also called antisense therapy. Since the HGP, development of DNA-based therapeutics, thereby, has taken center stage.

In the last few years, there has been an outbreak of studies investigating DNA-based therapeutics. A gradual evolution of gene medicine from an experimental technology into a viable strategy is taking place. The foremost advantage of using DNA-based pharmaceuticals over conventional drugs is their selective recognition of targets and pathways. This is due to the high specificity and selectivity of interaction with the targets and theoretically, has reduced potential for toxicity, fewer side effects and, therefore, safer drugs. Having said that, these DNA-based compounds are new drug candidates and their safety with respect to long-term use needs to be determined. Moreover, these DNA-based molecules are generally characterized by poor cellular uptake and rapid degradation. This necessitates the use of delivery vectors for efficient internalization and activity.

Antisense therapy, a type of gene therapy, is a technique for gene silencing whereby expression of an aberrant gene is silenced using short stretches of chemically modified nucleotides known as oligonucleotides. The technique is given this name because the synthesized oligonucleotide has a base sequence that is complementary to the gene's messenger RNA (mRNA), which is called the "sense" sequence. The antisense oligonucleotide—either a DNA or RNA molecule—thus binds to the mRNA produced by the aberrantly expressing gene and inactivates or degrades it.

Unmodified oligonucleotides are cheap and uncomplicated to synthesize. However, their use is limited as they are rapidly degraded. Moreover, their degradation products may be toxic and damaging to cell growth and multiplication. The general strategy to mediate gene silencing is to introduce certain modifications within the oligonucleotides and thereby use them in isolation or in combination in a single oligonucleotide. Extensive research has led to the development of a range of modifications of antisense oligonucleotides (ASOs), the most effective ones being morpholinos, peptide nucleic acids (PNAs), aptamers, 2'-O-methyl oligonucleotides, and phosphorothioates (Chan, Lim, and Wong 2006).

Phosphorothioates—first-generation ASOs—are the most widely studied and utilized oligonucleotide modifications because of their high stability and relative ease of synthesis. They are designed by modifying the backbone structure of DNA molecules. Phosphorothioates have excellent antisense activity and cause silencing by blocking the synthesis of proteins from target mRNA or by degrading them. In spite of being the most popular oligonucleotides, phosphorothioates do tend to display suboptimal and independent off-target effects in vivo. In order to overcome these drawbacks, newer modifications were developed to generate the second generation ASOs (Chan, Lim, and Wong 2006). For instance, replacement of a hydrogen atom at a certain position gave rise to 2'-O-methyl oligonucleotides that are characterized by higher stability. Furthermore, modifications by replacing nucleic acids with their analogues gave rise to extremely stable and high-affinity molecules called PNAs. The PNAs bring about an antisense effect by virtue of steric hindrance; moreover, they are resistant to nucleases—enzymes that degrade nucleotides—as well. Furthermore, another kind of nucleic acid analogue was developed —morpholinos— which has also become popular in recent years for functional studies. Morpholinos block the access of other molecules to short stretches of RNA, thereby preventing cell from making that protein. PNAs, morpholinos, and phosphoroamidates (another kind of backbone modification) constitute the third-generation ASOs.

More recently, single- or double-stranded molecules have been designed that can bind directly to the proteins. These nucleic acid segments—called aptamers (Sun et al. 2014) interfere with the molecular functioning of the target proteins. Their molecular recognition properties are equivalent to, or even better than, antibodies; however, they have an added advantage of being more specific (and less immunogenic) as they are engineered through repeated rounds of selection. This selection process is called SELEX (systematic enrichment of ligands by exponential enrichment) and can be carried out completely in a test tube. In spite of these characteristics, unmodified aptamers have stability issues.

Another strategy to implement DNA-based therapeutics is by the use of large, double-stranded DNA molecules called plasmids. Plasmids are artificial constructs that code for the therapeutic entity, that is, specific proteins, as opposed to antisense therapy that leads to silencing of such proteins (Patil, Rhodes, and Burgess 2005). These plasmids are internalized by the cell and by utilizing the cellular machinery; they encode functionally competent copies of a protein and hence correct genetic errors. In addition to disease treatment, plasmids can be used as DNA vaccines for genetic immunization as well.

Despite this wide array of options and advantages, one of the principal roadblocks in DNA-based therapeutics is their delivery to the target site within the body. In order to bring about change in gene expression, penetration into the cells is of paramount importance; the mechanism of the process, however, is still unclear. Moreover, for treatment of brain ailments, crossing the blood-brain barrier is another challenge considering its highly selective nature. As a means to overcome this issue, vectors have been designed that are made up of cationic polymers or lipids that act as carriers for these molecules and help in their internalization by the cells. These carriers have shown variable degrees of success. Strikingly, in some in vivo experiments, even naked DNA molecules have shown high degrees of efficacy (Achenbach, Brunner, and Heermeier 2003). The other challenge in the use of DNA-based molecules would be their inevitable toxicity. Although antisense technology is engineered to be very specific, it can still cause unintended damage and modify gene expression in an unrelated tissue.

Until 2015, fomivirsen (trade name—Vitravene) and mipomersen (trade name—Kynamro) were the only two antisense drugs to have been approved by the FDA (Orr 2001). Fomivirsen is used to treat an eye disease associated with AIDS, and mipomersen (Furtado, Wedel, and Sacks 2012) is a cholesterol-reducing drug used to treat familial hypercholesterolemia. Both these drugs have been developed by ISIS Pharmaceuticals (now Ionis Pharmaceuticals). Recent advances in aptamer technology have also led to the development of Macugen8 (Vinores 2006) (co-developed by Eyetech and Pfizer). It was approved recently by the U.S. FDA and used for the treatment of age-related macular degeneration, a disorder characterized by excessive pathological blood vessel growth at the center of the retina.

Over the recent past, with the rapid advancements in technology and our knowledge of the cellular system, DNA-based therapeutic technology has emerged as a valid approach to selectively modulate gene expression. A colossal amount of research is being carried out by numerous groups around the world to tap into this technology. Development of treatments for diseases like cystic fibrosis, Duchenne muscular dystrophy, spinal muscular dystrophy, Huntington's disease, hemophilia, and sickle-cell disease, to name a few, is already in advanced stages of research (Martinez et al. 2013). Owing to its vast potential, the number of experiments has increased continuously, leading to the development of novel therapeutic compounds, a lot of which appear to be preliminarily positive. However, the optimal use of DNA-based drugs is still marred with problems relating to effective design, enhanced biological activity, and efficient target delivery. These issues are currently being actively addressed and will hopefully continue to shed light on ways to increase therapeutic efficacy and specificity.

References

Achenbach, T. V., B. Brunner, B. and K. Heermeier, K. 2003. "Oligonucleotide-Based Knockdown Technologies: Antisense versus RNA Interference." *ChemBioChem.* 4: 928–935.

Chan, J. H. P., S. Lim, and W. S. F. Wong. 2006. "Antisense Oligonucleotides: From Design to Therapeutic Application." *Clinical and Experimental Pharmacology and Physiology.* 33: 533–540.

Furtado, J. D., M. K. Wedel, and F. M. Sacks. 2012. "Antisense Inhibition of ApoB Synthesis with Mipomersen Reduces Plasma ApoC-III and ApoC-III-Containing Lipoproteins. *Journal of Lipid Research.* 53: 784–791.

Martinez, T., et al. 2013. "Silencing Human Genetic Diseases with Oligonucleotide-Based Therapies." *Human Genetics.* 132: 481–493.

Orr, R. M. 2001. "Technology Evaluation: Fomivirsen, Isis Pharmaceuticals Inc/CIBA Vision." 2001. *Current Opinion in Molecular Therapies.* 3: 288–294.

Patil, S. D., D. G. Rhodes, and D. J. Burgess. 2005. "DNA-Based Therapeutics and DNA Delivery Systems: A Comprehensive Review. *AAPS Journal.* 7: E61–E77.

Sun, H. et al. 2014. "Oligonucleotide Aptamers: New Tools for Targeted Cancer Therapy." *Molecular Therapy. Nucleic Acids.* 3: e182 (2014).

Venter, J. C. et al. 2001. "The Sequence of the Human Genome." *Science.* 291(5507): 1304–1351.

Vinores, S. A. 2006. "Pegaptanib in the Treatment of Wet, Age-Related Macular Degeneration." *International Journal of Nanomedicine.* 1: 263–268 (2006).

Manish Muhuri finished his doctorate from Nanyang Technological University, Singapore, in RNA biology while working on molecular mechanisms of brain development in mice. Currently, he is working as a Research Fellow in the Institute of Medical Biology, Singapore.

CRISPR and Beyond—What the Future Holds for Gene Editing
Sheila T. Yong

Imagine DNA as a giant computer that stores tremendous amounts of data. Just as lines of code instruct a computer to perform certain functions, data stored in DNA contain instructions for building the biomolecules required to sustain the processes of life in all organisms.

We can edit a computer program's code by adding, deleting, and replacing parts of the code to yield a desired output. Analogously, gene editing is the act of changing the data stored in DNA to alter the properties of a living organism.

In practice, gene editing is not a new concept. In fact, it began even before humans knew what DNA was. For centuries, humans have been cultivating crops and breeding livestock over many generations, systematically carrying out planned crosses in an effort to propagate plants and animals with the same desirable traits generation after generation. For instance, the corn we eat today looks nothing like its wild counterpart, which was domesticated by indigenous Mexican people about 10,000 years ago ("The Evolution of Corn" 2016). At least 150 breeds of dogs exist today, courtesy of interbreeding over a period of at least 150 years that began with wild wolves ("Evolution of the Dog" 2001). The comparison of the genomes of these domesticated plants and animals with those of their wild ancestors reveals significant alterations in their genomes; they now embody a rewritten program for making bigger, tastier corn, or dogs that are loyal companions instead of pure hunters. A human guiding hand has made the selection process less random, and directed it toward a defined purpose.

Currently taking the center stage of the gene-editing debate is the CRISPR/Cas9 system, which, like gene editing, is not a new development. The system's genetic components have existed in nature for at least 2.5 billion years as they form the adaptive immune system of archaea and bacteria (Gribaldo and Brochier-Armanet 2006). CRISPRs are small segments of foreign DNA incorporated into bacterial and archaean genomes as a result of an earlier infection (Makarova et al. 2011). CRISPRs, together with the Cas9 endonuclease, create a "memory" to protect these microorganisms from future invasion by the same pathogens.

What is new about CRISPR/Cas9 is that scientists have found a way to harvest its potential. They have created a modified version that uses a single guide RNA to lead the Cas9 enzyme to the targeted DNA sequence in the genome and cut the DNA at a specific site (Jinek et al. 2013). The cleaved DNA can then be repaired using an exogenous DNA template with predetermined nucleotide sequences.

In fact, altering the genome at the cellular level is not a new concept either. In 1928, Frederick Griffith successfully transformed an avirulent strain of *Streptococcus pneumoniae* into a virulent strain by simply mixing dead cells of the latter with live avirulent cells (O'Connor 2008). Since then, scientists have developed various gene-editing methods to generate organisms that express the desired traits. However, many of these procedures are difficult and tedious, and are often accompanied by errors and unexpected outcomes.

The advantage of CRISPR/Cas9 over conventional gene-editing methods is its precision and ease of use because its mechanism of incorporating the desired DNA sequence into the genome is much simpler. More importantly, it works in virtually any cell type, including embryos (Yin 2015). In fact, scientists in China are using CRISPR extensively to modify embryos from various animals, including goats, pigs, monkeys and dogs (Symons 2015).

Unfortunately, the power of CRISPR/Cas9 is also its downfall. The major concern is that this technology may be used to engineer human embryos to produce designer babies and generate inheritable genetic changes.

This worry is not unfounded. Two Chinese research groups have used CRISPR/Cas9 to engineer human embryos to rectify a mutation that causes a fatal blood disorder (Cyranoski and Reardon 2015), and to modify the CCR5 gene in immune cells to render them resistant to HIV infection (Pascual 2016), respectively. Along these lines, will it be possible to custom-design babies? Can we create a real-life Captain America–type "super soldier"?

Before we indulge in such possibilities, we must understand that many human traits are controlled by multiple genes, and the full extent of this control remains unknown. Therefore, it is impossible to manipulate every single gene involved to the exact specification to yield the desired traits. Moreover, despite its precision, CRISPR/Cas9 can still have off-target effects and generate unwanted mutations. Out of the 26 embryos targeted for CCR5 gene editing, only four were genetically modified and not all chromosomes incorporated the desired mutation (Pascual 2016).

Like any technology, CRISPR/Cas9 will continue to evolve as researchers uncover ways to improve the procedure. For instance, they have discovered a Cas9 enzyme variant that is smaller in size and thus can be delivered more easily into human cells (Ledford 2015). More recently, a new CRISPR/Cas9 gene-editing protocol that does not involve the use of foreign DNA has been reported (Williams 2016). As scientists continue to learn about CRISPR/Cas9, they will devise ways to help this technology reach its full potential. That said, should CRISPR/Cas9 take all the blame for the impact it may have?

No matter how advanced CRISPR/Cas9 eventually becomes, it is still a tool subject to human use and manipulation. As with all new technologies, there will always be a struggle between the discovery of new knowledge and how such knowledge is applied or regulated. As Spiderman's Uncle Ben famously proclaimed, "With great power comes great responsibility." The technology itself is not at fault. It is up to us to decide how we want to use it.

References

Cyranoski, David, and Sara Reardon. 2015. "Chinese Scientists Genetically Modify Human Embryos." http://www.nature.com/news/chinese-scientists-genetically-modify-human-embryos-1.17378. Accessed on April 10, 2016.

"The Evolution of Corn." 2016. Learn.Genetics. http://learn.genetics.utah.edu/content/selection/corn/. Accessed on April 20, 2016.

"Evolution of the Dog." 2001. Evolution. http://www.pbs.org/wgbh/evolution/library/01/5/l_015_02.html. Accessed on April 22, 2016.

Gribaldo, Simonetta, and Celine Brochier-Armanet. 2006. "The Origin and Evolution of Archaea: A State of the Art." *Philosophical Transactions of the Royal Society B: Biological*

Sciences. doi: 10.1098/rstb.2006.1841. Accessed on April 21, 2016.

Jinek, Martin, et al. 2013. "RNA-Programmed Genome Editing in Human Cells." *Elife*. doi:10.7554/eLife .00471. Accessed on May 6, 2016.

Ledford, Heidi. 2015. "Mini Enzyme Moves Gene Editing Closer to the Clinic." http://www.nature.com/news/mini-enzyme-moves-gene-editing-closer-to-the-clinic-1.17234. Accessed on April 10, 2016.

Makarova, Kira S., et al. 2011. "Evolution and Classification of the CRISPR-Cas System." *Nature Reviews Microbiology*. http://www.ncbi.nlm.nih.gov/pmc/articles/PMC3380444/. Accessed April 21, 2016.

O'Connor, Claire. 2008. "Isolating Hereditary Material: Frederick Griffith, Oswald Avery, Alfred Hershey, and Martha Chase." Nature Education. http://www.nature.com/scitable/topicpage/isolating-hereditary-material-frederick-griffith-oswald-avery-336. Accessed on May 9, 2016.

Pascual, Katrina. 2016. "Chinese Scientists Edit Genes of Human Embryos for the Second Time." http://www .techtimes.com/articles/149281/20160412/chinese-scientists-edit-genes-of-human-embryos-for-the-second-time.htm. Accessed on April 18, 2016.

Symons, Xavier. 2015. "CRISPR-Mania in China." http:// www.bioedge.org/bioethics/crispr-mania-in-china/ 11670. Accessed on April 10, 2016.

Williams, Ruth. 2016. "Gene Editing without Foreign DNA." http://www.the-scientist.com/?articles.view/ articleNo/45155/title/Gene-Editing-Without-Foreign-DNA/. Accessed on April 22, 2016.

Yin, Steph. 2015. "What Is CRISPR/Cas9 and Why Is It Suddenly Everywhere?" http://motherboard.vice.com/ read/what-is-crisprcas9-and-why-is-it-suddenly-everywhere. Accessed on April 10, 2016.

Sheila T. Yong holds a PhD in molecular cancer biology from Duke University. She is the owner and founder of Hornbill Scientific Co., a science writing and consulting business that helps scientists communicate their research effectively to a wide variety of audiences. Her goal is to make the knowledge of science accessible to experts and lay audiences alike, by bridging the gaps among researchers in different disciplines as well as the gap between researchers and the general public.

Forensic Use of DNA Technology
Jon Zonderman

In mid-April 2016, Keith Allen Harward walked out of prison in Burkeville, Virginia, after serving 33 years for a murder and a rape he did not commit. As with hundreds of cases since the mid-1990s, DNA technology was used to clear him. Mr. Harward had been convicted of beating a man to death and raping his wife during a 1982 home invasion in Newport News, Virginia, while in the navy serving on an aircraft carrier docked there.

At the time, police had no firm suspects and little evidence. Harward was convicted on the eyewitness testimony of a navy guard who identified him as the young sailor who returned to base after the crime had been committed with his uniform splattered with blood or paint. Prosecutors tied this identification to the victim's testimony, even though she said the sailor who raped her had been clean-shaven and Harward had a mustache at the time.

The key piece of evidence, however, was "bite mark" testimony by two forensic dentists, who testified that Harward's tooth pattern almost certainly matched that of the man who had bitten the victim during the assault. In 2009, the National Academy of Sciences said bite-mark analysis, as well as tool-mark analysis, tire-tread analysis, footprint analysis and voice analysis, had never been scientifically validated, although all of these techniques had been used in criminal cases for decades. Analysis of an item in the rape kit from the 1982 case, a towel used to cover the woman after the rape, showed male DNA from semen to be that of Harward's shipmate Jerry Crotty, who died in 2006 while serving prison time for other crimes.

This case in many ways highlights the "double-edged" aspect of using DNA for forensic analysis in crime solving and criminal proceedings. I first wrote about the forensic use of DNA technology, as well as bite-mark analysis, in my book *Beyond the Crime Lab: The New Science of Investigation* (1990, second edition 1998, John Wiley & Sons).

DNA analysis, often referred to as "genetic fingerprinting," was first used in the United States in 1987 in a Florida rape case. Tommie Lee Andrews, a 24-year-old Orlando pharmaceutical company employee, was convicted through comparison of male DNA on a vaginal swab taken from the rape victim and a sample of his blood obtained under court order. A private New York laboratory, Lifecodes, performed the analysis using its trademarked "DNA-print" test.

In every state in which the technology was introduced, a court hearing was held to determine if the technology was well-enough established in the scientific community for courts to consider it valid. Within a couple of years most state courts and the federal courts had accepted the technology and the FBI began performing DNA analysis in its national laboratory. Many state police labs and the labs of police departments in large municipalities followed.

While huge pieces of DNA are the same in all people, small sections are specific to one person per many hundreds of thousands, if not one in a million. The technology used in criminal investigations—restriction fragment length polymorphis (RFLP)—is the same as used in paternity testing, which is what Lifecodes was mostly involved with before police and prosecutors asked for help. While saying that a DNA sample is specific to one person in hundreds of thousands is not proof positive, combined with other evidence it is as powerful or more powerful as traditional fingerprints and far more specific than use of traditional blood type statistical analysis.

In the early years of using DNA analysis in criminal cases, some prosecutors played fast and loose. In one case in Connecticut in 1990, before the state had established its own DNA testing procedures in its state police lab, prosecutors asked the FBI to analyze DNA samples in a rape case. An FBI analyst testified that DNA samples taken from the victim's underwear and the rape kit were not those of the defendant, Ricky Hammond. Despite this, the prosecutor had a credible victim, who recalled the perpetrator's appearance and even gestures and tics. Hammond was convicted and served over two years before a series of appeals, dismissals, and retrials, the final of which resulted in his plea to a charge of unlawful restraint and a sentence of the time he had already served. At the time of the original conviction, I wrote an op-ed column for the *Hartford Courant* taking prosecutors to task for using DNA to convict one day and then denying its validity at the next trial when the DNA exonerated a defendant.

It was not long after, 1992, that two New York attorneys, Barry Scheck and Peter Neufeld, began The Innocence Project, which uses DNA technology to reinvestigate cases where prisoners claim wrongful conviction by eye-witness testimony, circumstantial evidence, or the analytical techniques disqualified in the 2009 study. To date, the organization and similar organizations run from law schools and journalism schools around the country have helped free hundreds of Americans.

There are three key issues today that we should be concerned about.

First, there are literally thousands of rape kits across the country that have not yet been analyzed, usually for lack of funds and personnel. This means that guilty people are still free to harm others, but also that innocent people continue to be imprisoned. Some private foundations are funding efforts to cut these backlogs, but taxpayers must also step up to the plate.

Second, police investigators are increasingly asking ancestry registry organizations, like Ancestry.com and 23andMe, to search their databases for potential matches to crime scene or suspect samples. Remember, DNA typing provides a match in only one of a few hundreds of thousands of individuals. Putting that together with who lives in one city can be powerful, but your DNA may be a very close match to someone else's who lives thousands of miles from you. Should these companies turn over private information people have given them for a different purpose?

Third, many states enacted mandatory blood sample laws for those arrested and/or convicted of crimes. Many Americans are arrested but charges are dismissed or they are found innocent at trial. Should their DNA be in a state or national database, or only those of people actually convicted?

Jon Zonderman, author of Beyond the Crime Lab, *is co-author with his wife, a pediatrician, of six books about health science. He has written hundreds of articles on science, technology, and business, and is a lecturer in journalism at Southern Connecticut State University, where he teaches magazine writing, literary journalism, and other classes.*

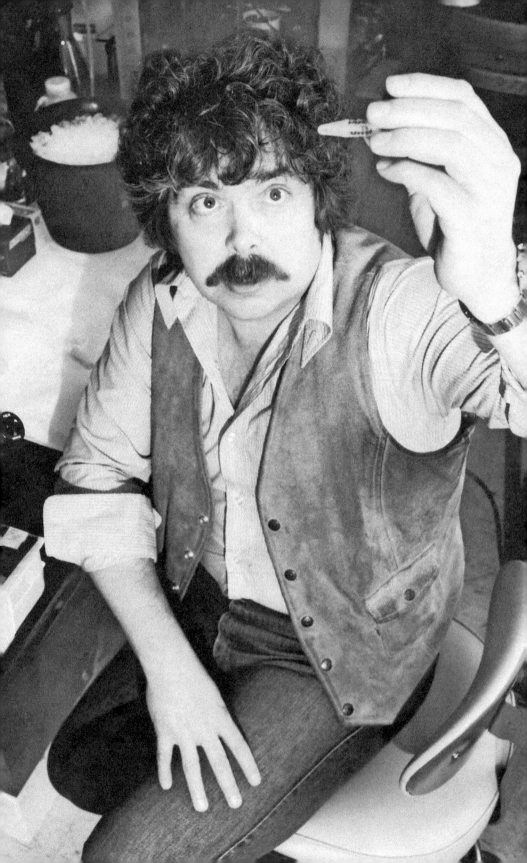

4 Profiles

Introduction

One useful and interesting way to learn more about the history and background of DNA technology is to review brief sketches of some of the most important individuals and organizations that have been involved in that story. In the brief space available here, one can include only a sample of those individuals and organizations who have made DNA technology what it is today. These sketches may provide a good starting point for the reader who would like to learn more about some of the pioneers in the field of DNA technology.

Accreditation Council for Genetic Counseling (ACGC)

The Accreditation Council for Genetic Counseling is a spin-off of the American Board of Genetic Counseling. ACGC was created in 2011 to take on the responsibility for establishing standards for training programs for genetic counselors and for certifying programs that meet those standards. The council began operation on January 1, 2013, and is responsible for the accreditation of about three dozen programs in 22 states in the United States and three Canadian provinces.

Biochemist Dr. Herbert Boyer holds a vial while performing an experiment at his University of California lab. (Steve Northup/The LIFE Images Collection/Getty Images)

AgBioWorld

AgBioWorld was founded in January 2000 by Professor C.S. Prakash of Tuskegee University and Gregory Conko of the Competitive Enterprise Institute with the goal of providing reliable scientific information on the use of biotechnology in agriculture to all concerned parties throughout the world. The organization is a 501(c)(3) nonprofit organization that depends primarily on volunteers for its operation. It attempts to remain rigorously unbiased in its work, rejecting contributions of any organization with a vested interest in favor of or opposed to the use of biotechnology in agricultural operations. The major activity of AgBioWorld is providing information on agricultural biotechnology. In concert with that activity, the organization also provides referrals to experts in the field who can be contacted for additional, detailed information on a variety of topics, such as insect- and herbicide-tolerant crops, GM food safety and human health, feed safety and animal health, economic issues related to biotechnology, risk assessment and public perception of biotechnology, plant biology, breeding and development, food and trade policy, regulations and legal issues, and liability for GMOs and non-GMOs. AgBioWorld has also written a Declaration of Support for Agricultural Biotechnology that has now been signed by 3,400 scientists worldwide, including 25 Nobel Prize winners.

American Board of Genetic Counseling (ABGC)

The American Board of Genetic Counseling is the accrediting agency for genetic counseling in the United States and Canada. It establishes standards in the field by the accreditation of graduate programs in genetic counseling and conducting examinations that lead to diplomates as certified genetic counselors. ABGC was first incorporated in 1993 for the purpose of accrediting individuals in the field of genetic counseling, as well as accrediting programs of training in genetic counseling. In 2013, the latter function was spun off to a new and separate organization, the Accreditation Council for Genetic Counseling (ACGC).

American Society of Bioethics and Humanities (ASBH)

The American Society of Bioethics and Humanities was formed in 1998 in the consolidation of three existing organizations: the Society for Health and Human Values, the Society for Bioethics Consultation, and the American Association of Bioethics. The mission of the organization is to promote the exchange of ideas and to foster multidisciplinary, interdisciplinary, and interprofessional scholarship, research, teaching, policy development, professional development, and collegiality among people engaged in clinical and academic bioethics and the medical humanities. ASBH conducts an annual meeting, a spring meeting, and an annual National Undergraduate Bioethics Conference, as well as endorsing and publicizing a number of other meetings by sister societies with interests and concerns compatible with its own. The organization also provides an online job bank for members.

American Society of Gene and Cell Therapy (ASGCT)

The American Society of Gene and Cell Therapy is a professional association of scientists and researchers working for the development of new technologies for gene and cell therapy. The organization was founded in 1996 with fewer than 1,000 members. It has grown to twice that size, with the largest number of members (about 1,300) from the United States. The association is interested in promoting research on new gene and cell therapies, providing opportunities for the exchange of information among researchers in the field, and promoting the development of methods by which research findings can be translated into clinical studies. In addition to its support for professional activities in gene and cell therapy, the society provides educational materials to the general public. The two categories of membership are traditional or active members, consisting of individuals who are currently working in the field of gene or cell therapy; and associate members, consisting of graduate students and postdoctoral researchers.

American Society of Human Genetics (ASHG)

The American Society of Human Genetics was founded in 1948 as the primary organization of human geneticists world-wide. Today, its 8,000 members include researchers, academicians, clinicians, laboratory practice professionals, genetic counselors, nurses, and other individuals with a special interest in the field of human genetics. The organization has a four-fold list of objectives that includes sharing of recent research results through an annual convention and the association's professional journal; advocating for research support in the field of human genetics; providing educational materials for professionals and the general public; and promoting and supporting responsible scientific and social policies related to human genetics. The organization carries out its work through eight committees: awards, executive, finance, information and education, nominating, program, social issues, and professional development.

W. French Anderson (1936–)

Anderson is sometimes called the father of human gene therapy because of his role in some of the earliest efforts to cure disease by gene transfer methods. In 1990, he treated a four-year-old girl with severe combined immunodeficiency disease (SCID), which occurs when a genetic error prevents a child's immune system from developing normally. As a result, the child is highly susceptible to all forms of infection and, often, dies at an early age. Anderson and his colleagues proposed to cure the disease by introducing functioning copies of the defective genes into the child's genome, restoring its ability to produce healthy components of the immune system.

William French Anderson was born in Tulsa, Oklahoma, on December 31, 1936. He developed an avid interest in science and mathematics at an early age, but was also interested in a variety of extracurricular activities, including theater, track, and debate. He is said to have become interested in the last of these subjects because of his tendency to stutter, a problem he thought he could resolve with demanding attention to speaking. He also later expressed an interest in studying the genetic basis for stuttering.

Anderson entered Harvard College in 1954, where he was exposed to the most recent advances in genetics, especially in a course taught by James Watson. The course was instrumental in convincing him to change his college major from mathematics to genetics. After receiving his BS from Harvard in 1958, he went to England to continue his studies under Francis Crick at Cambridge University. He earned his MA from Cambridge in 1960, and then returned to Harvard to study medicine, earning his MD in 1963.

In 1965, Anderson took a research position with the National Institutes of Health (NIH), where he spent the next 30 years working on methods for treating genetic disorders with engineered cells. He chose to focus on a disease known as adenosine deaminase deficiency disease (ADA) because it develops as the result of a single defective gene, a much simpler condition than many other types of genetic disease. ADA is one of the conditions commonly associated with SCIS. Anderson extracted T cells from the four-year-old patient, inserted correct copies of the ADA gene into those cells, and then returned the engineered cells to her body. He was able to show that the cells survived, reproduced, and began to synthesize the adenosine deaminase previously absent from her immune system.

In 1992, Anderson left the NIH to accept an appointment as professor of biochemistry and pediatrics and director of the Gene Therapy Laboratories at the Keck School of Medicine at the University of Southern California. In 2004, he was charged with sexual abuse of a minor, convicted of the offense, and sentenced to 14 years in prison. In 2014, Anderson created a Web site containing documents that he claimed would support his innocence in the case (www.wfrenchanderson.org).

Werner Arber (1929–)

Arber discovered an important family of chemical compounds known as restriction enzymes that cut DNA molecules at specific points in their polynucleotide chains. Prior to Arber's work, scientists had found that bacteria have evolved mechanisms for protecting themselves against attacks by viruses (called bacteriophages, or just phages), just as higher organisms have evolved immune systems to protect themselves against attack by viruses, bacteria, fungi, and other foreign organisms. Arber was able to show that the "immune system" used by bacteria consists of enzymes that cut the DNA of invading viruses, disabling them and preventing their continued infection of the bacteria. Arber's discovery was an important milestone in the development of DNA technologies, because it provided a method by which DNA molecules can be taken apart for the insertion of gene fragments from other sources, making possible the formation of recombinant DNA molecules.

Werner Arber was born in Gränichen, Switzerland, on June 3, 1929. After completing his public school education in 1945, he enrolled at the Kantonsschule Aarau gymnasium, from which he received a second-level "maturity" certificate in 1949. He continued his studies in chemistry and physics at the Swiss Polytechnical School in Zürich, where he earned his degree in 1953. Arber then chose to accept an assistantship in electron microscopy at the University of Geneva, a position in which he became more interested in and better informed about problems of bacteriophage physiology and genetics. He eventually decided to make a study of phages the focus of his career.

Arber was awarded his doctorate at Geneva in 1958 and then spent two years in a postdoctoral program at the University of Southern California in phage genetics. Toward the end of that program, he had an opportunity to visit the laboratories of a number of pioneers in that field, including those of Gunther Stent, Joshua Lederberg, and Salvador Luria, where he learned about the most recent advances in phage genetics and molecular biology. When Arber returned to the University of Geneva in 1960, he began the research that would lead to his discovery of restriction enzymes.

Arber was promoted to associate professor in molecular biology at Geneva in 1965 and, in 1971, he moved to the University of Basel as full professor of molecular microbiology, a post he held until his retirement in 1996. From 1986 to 1988, Arber also served as rector of the University of Basel. He was awarded a share of the 1978 Nobel Prize for Physiology or Medicine. Arber continues his affiliation with the University of Basel, where he is emeritus professor of molecular microbiology.

Armed Forces DNA Identification Laboratory (AFDIL)

The mission of the Armed Forces DNA Identification Laboratory is to provide research, consultation, and educational services to the U.S. Department of Defense and other federal agencies and to maintain and distribute, where needed, DNA specimens of U.S. military and other authorized personnel. The primary task of the agency is to identify members of the armed services who have been killed or lost in battle.

Association of Forensic DNA Analysts and Administrators (AFDAA)

The Association of Forensic DNA Analysts and Administrators is a nonprofit organization of individuals involved in the forensic and legal applications of DNA typing. The organization currently has about 200 members from more than

75 government agencies and private corporations in 25 state and international laboratories. The organization meets about twice a year at the Texas Department of Public Safety in Austin. The objectives of the organization are to help members stay up to date with DNA technology, promote dissemination of information about DNA typing to the professional community, discuss legislative issues relating to forensic DNA typing, provide an opportunity for professionals in the field to network with each other, and offer formal training in DNA typing and analysis to members.

Oswald Avery (1877–1955)

In 1944, Avery reported on a classic series of experiments conducted with postdoctoral colleagues Colin MacLeod and Macyln McCarty over a period of years. Those experiments were designed to identify the unit within living organisms responsible for the transmission of genetic traits. At the time, most scientists thought that some type of protein played this role. Heredity is a complex phenomenon that requires a complex molecule for transmission of traits to occur. Proteins certainly fit that role as they are large, complex molecules. In the Avery-MacLeod-McCarty experiments, laboratory mice were injected with two forms of pneumococcus (the bacterium that cause pneumonia), one that was virulent (disease-causing and fatal) and one that was not. By treating the mice with various combinations of bacteria, the researchers found that the compound responsible for transmitting resistance to pneumococci in mice was DNA, not a protein. This research is of such profound significance in the history of DNA that Avery is sometimes described as the most deserving researcher never to have received a Nobel Prize.

Oswald Theodore Avery was born on October 21, 1877, in Halifax, Nova Scotia. He moved with his family to New York City in 1887, where he attended the New York Male Grammar School. He continued his studies at the Colgate Academy, preparatory school of Colgate University, and then at the university itself. He earned his BA in humanities in 1900, but then chose to study medicine at the Columbia University College of Physicians and Surgeons in New York, where he received his MD in 1904.

Avery remained in practice for only a relatively short time, primarily because he was frustrated at his inability to help some of his patients. He decided instead to turn to a career in research and, in 1907, accepted a position as associate director of the division of bacteriology at the Hoagland Laboratory in Brooklyn, New York. Of the many research papers he wrote at the Hoagland, one caught the eye of Rufus Cole, then director of the Hospital of the Rockefeller Institute for Medical Research. Impressed by Avery's work, Cole offered him a position at Rockefeller in 1913. He remained at Rockefeller until he retired from research in 1948. Although he was given emeritus status by the university in 1943, Avery continued his research there for another five years. It was near the end of his active involvement in scientific research, then, that he, along with MacLeod and McCarty, made his revolutionary discovery about DNA. In 1948, Avery left Rockefeller and moved to Nashville, Tennessee, to be near his brother. He died in Nashville on February 2, 1955, of pancreatic cancer.

Paul Berg (1926–)

In the late 1960s, Berg made the first recombinant DNA (rDNA) molecules by splicing together foreign genes from a variety of sources into the DNA molecules of host organisms, such as bacteria and mammalian cells. This achievement not only brought together a number of techniques previously developed to produce recombinant organisms, but also provided a method for studying the characteristics and outputs of specific segments of DNA, those inserted into host organisms.

Paul Berg was born in New York City on June 30, 1926. He attended Pennsylvania State University, from which he received his BS in biology in 1948, and Western Reserve University (now Case Western Reserve University), where he was awarded his PhD in biochemistry in 1952. After four years of postdoctoral research at the NIH, the Institute of Cytophysiology in Copenhagen, and the Washington University School of Medicine in St. Louis, Berg accepted an appointment as assistant professor of microbiology at the Washington University School of Medicine. In 1959, Berg moved to Stanford University as associate professor in the School of Medicine. He spent the rest of his academic career at Stanford in a number of positions, including Sam, Lulu, and Jack Willson Professor of Biochemistry; Vivian K. and Robert W. Cahill Professor in Biochemistry and Cancer Research; Director, Beckman Center for Molecular and Genetic Medicine; and (currently) Cahill Professor in Biochemistry, Emeritus, and Director of the Beckman Center for Molecular and Genetic Medicine, Emeritus. He was elected to the National Academy of Sciences in 1966 and was awarded the Nobel Prize in Chemistry in 1980 for his research on recombinant DNA molecules.

After announcing the results of his research on recombinant DNA, Berg became concerned about possible safety, social, and ethical implications of such research. He wrote to the NIH, outlining these concerns, and encouraged his fellow scientists to meet and consider the possible risks involved in rDNA research. Largely because of Berg's efforts, a group of more than 100 researchers, legal experts, and others concerned about rDNA research met at the Asilomar Conference Center in California in 1975 to work out guidelines for the monitoring and control of such research. Those guidelines were the first regulations ever developed for research on DNA technology.

Biotechnology Innovation Organization (BIO)

The Biotechnology Innovation Organization (BIO) was formed in 1993 by the merger of the Association of Biotechnology Companies and the Industrial Biotechnology Association to provide a single voice for companies working in the field of biotechnology to speak about regulation of biotech crops, small business and economic development issues, national health care policy, and reform of the FDA. At the time BIO was known as the Biotechnology Industry Organization. It changed its name to Biotechnology Innovation Organization in 2016 to better represent members' goal of providing innovative solutions to ways of helping to heal, feed, and fuel the world's population using biotechnology. BIO claims a number of successes in it more than 20 years of existence, including enactment of reforms in the FDA in the 1997 Food and Drug Administration Modernization Act, adoption of BIO principles on coverage of outpatient drug bills in the Medicare Modernization Act of 2003, dramatic growth in the amount of biotech crops planted each year, adoption by at least 40 states of programs to encourage biotech investment within the state, significant increases in farm income as a result of the use of biotech crops, and the cloning of endangered species using biotech technology.

Biotechnology Regulatory Services (BRS)

Biotechnology Regulatory Services is a division of the Animal Plant and Animal Health Inspection Service (APAHIS) of the U.S. Department of Agriculture (USDA). The regulation of genetically modified organisms has been an issue of concern to the USDA since 1978, when the National Institutes of Health issued the first federal guidelines for the monitoring of genetically engineered organisms in the United States. Regulatory responsibility was assigned to a series of committees and working groups until 2002, when BRS was formed with exclusive responsibility for issuing permits, conducting inspections, and carrying out other activities required for the monitoring of genetically engineered organisms in agriculture. In 2008, the bureau published a comprehensive plan for the future of biotechnology in agriculture and its regulation in the United States.

Herbert Boyer (1936–)

In the history of DNA technology, the names of Herbert Boyer and Stanley N. Cohen will forever be paired with each other. In 1972, the two researchers were both attending a conference in Hawaii on plasmids, circular loops of DNA found in bacteria and protozoa. Over lunch, the two men found that they were engaged in very similar research projects. Cohen was studying the antibiotic properties of certain bacterial plasmids, while Boyer was studying methods for introducing segments of DNA into precisely defined segments with ends that could be attached to other pieces of DNA. The two decided to collaborate and, within four months, had carried out one of the classic studies in the history of molecular biology. In this study, they introduced specified pieces of DNA into a bacterial plasmid (using methods developed by Boyer), and then inserted the plasmid into bacteria (using methods developed by Cohen). The results of these experiments were bacteria whose DNA contained clearly defined segments of foreign DNA (genes) capable of synthesizing specific proteins. When those bacteria reproduced, they then became tiny "factories" for the production of those proteins, the earliest forerunners of contemporary industrial rDNA technologies.

Herbert Wayne Boyer was born in Pittsburgh, Pennsylvania, on July 10, 1936. He received his AB from St. Francis College in 1968, his MS and PhD (bacteriology) from the University of Pittsburgh in 1960 and 1963, respectively. He did his postdoctoral work at Yale University as a U.S. Public Health Service research fellow. In 1966, Boyer left Yale to accept a position at the University of California at San Francisco (UCSF) as assistant professor of microbiology. He has remained with UCSF ever since and currently holds the title of emeritus professor of biochemistry and biophysics in the UCSF School of Medicine.

In 1976, Boyer and venture capitalist Robert A. Swanson founded the world's first corporation for the development of commercial products made with rDNA research, Genentech,

Inc. Within its first year of operation, Genentech had produced the first commercial rDNA product, the hormone somatostatin.

Among the honors given to Boyer include the National Medal of Technology, National Medal of Science, Biotechnology Heritage Award, Shaw Prize in Life Science and Medicine, Lemelson-MIT Prize, and the Albany Medical Prize. The graduate faculty at UCSF has created the Herbert W. Boyer Program in Biological Sciences in recognition of Boyer's contribution to the university.

California Certified Organic Farmer (CCOF)

California Certified Organic Farmer (now known in almost all cases as CCOF) was founded in 1973 to provide a uniform, widely accepted system of certification for organic foods, one of the first such organizations created in North America. Today the organization consists of more than 1,600 food production and processing companies throughout Canada and the United States. Over the past four decades, CCOF has greatly expanded its activities and now sponsors conferences and other meetings, proposes and supports legislation, provides certification for organic products, supports educational programs and efforts, and organizes and coordinates political actions. Its efforts in the fields of genetic engineering and cloning have been focused on legislation monitoring and limiting genetically modified crops, support of local initiatives on genetic engineering, the supply of information on GM crops to farmers and the general public, and prevention of the use of clones in agriculture.

Center for Bioethics and Human Dignity (CBHD)

The Center for Bioethics and Human Dignity was founded in 1993 by a group of Christian bioethicists who were concerned that Christian perspectives were not being included in discussions of a number of important socioscientific issues. These individuals joined together to form a nonprofit 501(c)(3) educational institution to remedy this problem, a program that is currently located at Trinity International University in Deerfield, Illinois. The center currently focuses its work on a number of topics in the field of bioethics, including biotechnology, cloning, end of life, genetics, health care and clinical ethics, health and spirituality, alternative medicine, neuroethics, reproductive ethics, sanctity of life, and stem cell research. Center researchers prepare position papers, bibliographies, and other research and educational materials on these topics, most available online at the center's Web site. The center has also sponsored conferences on topics in bioethics, including Global Bioethics, Healthcare and the Common Good, and Extending Life: Setting the Agenda for the Ethics of Aging, Death, and Immortality.

Center for Food Safety (CFS)

The Center for Food Safety is a nonprofit advocacy membership organization founded in 1997 by the International Center for Technology Assessment. Its mission is to challenge potentially harmful food production technologies and to promote sustainable alternatives. The center uses a variety of methods to achieve its goal, including litigation, petitions for rule making, support for other agricultural and food safety organizations, grassroots organizing, public education, and media outreach. Its work is currently organized around 11 major issues: genetic engineering (GE foods, GE food labeling, GE fish, GE animals, GE insects, and GE Trees), pollinators and pesticides, food and climate, soil, seeds, animal factories, organic and beyond, animal cloning, aquaculture, food safety (food safety, food irradiation, rbGH, mad cow disease, and sewage sludge), and nanotechnology

Center for Genetics and Society (CGS)

The Center for Genetics and Society was established in August 2001 as an outgrowth of an earlier group, the Exploratory Initiative on the New Human Genetic Technologies, a project of the Public Media Center in San Francisco. The center's position is that new genetic technologies have significant potential for improving the health and welfare of all human beings, but that they can also be used for less-than-salubrious purposes by researchers and industry. It works to educate the general public about the risks and benefits of new genetic technologies by sponsoring conferences and briefings, publishing reports, and providing support for other groups interested in these issues. The center's Web site is a treasure chest of valuable information on a host of topics such as animal technologies, assisted reproduction, DNA forensics, egg retrieval, genetic selection, hybrids and chimeras, inheritable genetic modification, medical gene transfer, personal genomics, reproductive cloning, research cloning, sequencing and genomics, sex selection, stem cell research, and surrogacy. It also includes extensive sections on policies on the state, national, and international level; a variety of social issues, such as arts and culture, bioethics, biotech and pharma, eugenics, and media coverage; and perspectives on issues such as disability, environmentalism, human rights, and LGBTQI.

Center for Genomics and Public Health (CGPH)

The Center for Genomics and Public Health is affiliated with the School of Public Health and Community Medicine at the University of Washington, in Seattle. The mission of the center is to integrate new knowledge in genomics into public health practices. Its goal is to provide educational resources to state public health agencies and practitioners, insurance providers, and the general public and to develop ways for using genomics to improve the health of the general public. The center provides online training sessions and offers a summer institute on genetic testing and related issues.

Erwin Chargaff (1905–2002)

Chargaff is best known for his discovery of the relationship among nitrogen bases in a DNA molecule, a relationship now known as Chargaff's Rules. According to those rules, the number of adenine units in a DNA molecule is equal to the number of thymine units, and the number of cytosine units is equal to the number of guanine units. A second, and somewhat less well-known, rule is that the composition of DNA (i.e., the relative amounts of adenine, cytosine, guanine, and thymine) varies from organism to organism. Although Chargaff himself never realized the implications of his rules for the composition of the DNA molecule, those rules proved to be an essential clue for the elucidation of the structure of the DNA molecule by James Watson and Francis Crick in the early 1950s.

Erwin Chargaff was born in Czernowitz, Austria (now Chernovtsy, Ukraine), on August 11, 1905. He showed an early precocity in the study of languages (he eventually was able to speak and understand 15 languages) and considered a career in philology. He eventually settled on science, however, and attended the University of Vienna to pursue a degree in chemistry. He received his PhD from Vienna in 1928 and then accepted a position at Yale University as Milton Campbell Research Fellow in Organic Chemistry. Two years later he returned to Europe, where he worked as assistant in charge of chemistry for the Department of Bacteriology and Public Health at the University of Berlin and as research associate at the Institut Pasteur in Paris.

In 1935 he returned again to the United States, where he became a research associate in the Department of Biochemistry at Columbia University, where he remained for the rest of his academic career. He retired from Columbia in 1974 as professor emeritus. His retirement was accompanied by some ill feelings between himself and the university, and he transferred his research activities to Columbia's affiliate, Roosevelt Hospital, where he continued to work until 1992. He died in New York City on June 20, 2002, at the age of 96.

By the 1950s, Chargaff had become somewhat disenchanted with the direction being taken by molecular biologists. He said that researchers were "running riot and doing things that can never be justified." He believed that nature was far too complex for humans to think of it and treat it as some sort of machine that can be manipulated at their will. He even warned that the manipulation of genetic materials was an even greater threat to the human race than was nuclear power.

Chargaff received a number of honors and awards during his lifetime, including the Pasteur Medal, the Carl Neuberg Medal, the Société de Chimie Biologique Medal, the Charles Leopold Mayer Prize, the H.P. Heineken Prize, the Bertner Foundation Award, the Gregor Mendel Medal, the National Medal of Science, and the New York Academy of Medicine Medal.

Emmanuelle Charpentier (1968–)

One of the classic collaborations in science occurred at a scientific conference being held in Puerto Rico in 2011. At that conference, Emmanuelle Charpentier, then working at Sweden's Umeå University, and Jennifer Doudna, at the University of California, Berkeley, started talking about their common interest in discovering the mechanisms by which bacteria protected themselves from infection by viruses. Both had identified a molecular system known as CRISPR as the key element in that process, a system that activated an enzyme called Cas9 for the actual attack on viral DNA or RNA. They decided to collaborate on their studies, undeterred by the fact that their laboratories were apart by more than 5,300 miles. Out of that collaboration came one of the most significant discoveries in DNA technology, a process for gene editing, in the 21st century.

Emmanuelle Charpentier was born on December 11, 1968, in Juvisy-sur-Orge, France. She attended the University Pierre & Marie Curie, in Paris, from which she earned her bachelor's degree in biochemistry and her master's degree and PhD in microbiology in 1991, 1992, and 1995, respectively.

She completed her postdoctoral studies at the Pasteur Institute in Paris in 1995–1996 and the Rockefeller University in New York in 1996–1997. She then continued her studies in the United States at the New York University Medical Center (1997–1999), St. Jude Children's Research Hospital, in Memphis (1999), and the Skirball Institute of Biomolecular Medicine in New York (1999–2002).

In 2002, Charpentier returned to Europe, where she served successively as Lab Head and Guest Professor, Assistant Professor, and then Associate Professor at the Institute of Microbiology & Genetics of the University of Vienna (2002–2009); Lab Head and Associate Professor in the Department of Molecular Biology at Umeå (2009 to present); and Professor at the Helmholtz Centre for Infection Research in Braunschweig (2013 to present) and Alexander von Humboldt Professor at the Hannover Medical School in Hannover (2013 to present). In 2015, Charpentier was named Director of the Max Planck Institute for Infection Biology in the Department of Regulation in Infection Biology at Berlin.

Doudna and Charpentier's work has brought them an impressive number of honors and awards, usually in concert, including the Theodor Körner Prize for Science and Culture of the city of Vienna (2009); Göran Gustafsson Prize for Molecular Biology of the Royal Swedish Academy of Sciences (2014); Dr. Paul Janssen Award for Biomedical Research (2014); Jacob Heskel Gabbay Award (2014); Grand Prix Jean-Pierre Lecocq of the French Academy of Sciences (2014); Breakthrough Prize in Life Sciences (2015); International Society for Transgenic Technologies Prize (2015); Louis-Jeantet Prize for Medicine (2015); Ernst Jung Prize in Medicine (2015); Hansen Family Award (2015); Princess of Asturias Award (2015); Gruber Foundation International Prize in Genetics (2015); Umeå University Jubilee Award (2015); Carus Medal of the German National Academy of Science, Leopoldina (2015); Massry Prize (2015); Otto Warburg Medal (2016); L'Oréal-UNESCO "For Women in Science" Award (2016); Canada's Gairdner

International Award (2016); and the Leibniz Prize of the German Research Foundation (2016).

Mary-Dell Chilton (1939–)

Chilton is best known for her research on the process by which *Agrobacterium* bacteria infect tobacco plants. Her team showed that a bacterium is able to transmit its DNA into the host plant genome and that, furthermore, removal of disease-causing genes from the bacterial DNA does not affect this process. The information gained from these studies demonstrated the feasibility of producing transgenic plants by inserting foreign genes into the genome of a host plant, giving it the ability to produce proteins not typically associated with the native plant. The experiments conducted by Chilton's team have now become classics in the field of genetic engineering of plants.

Mary-Dell Chilton was born in Indianapolis, Indiana, on February 2, 1939. She was originally interested in astronomy and was a finalist in the 1956 Westinghouse Science Talent Search for building "a long telescope in a short tube." She planned to major in astronomy when she entered the University of Illinois at Champaign Urbana in 1956, but soon found that her instructors did not take her seriously because she was a woman. She briefly transferred to physics, but found that subject boring, so eventually decided to major in chemistry, earning her BS in that subject in 1960 and her PhD, also in chemistry, in 1967. Chilton then accepted an appointment at the University of Washington, in Seattle, where she remained until 1979. It was at the University of Washington that she conducted much of her initial work on the production of transgenic plants.

In 1979, Chilton moved to Washington University in St. Louis, where she continued her research on the genetic modification of tobacco plants. After four years, she resigned her academic position in order to take a job with the CIBA-Geigy Corporation (now Syngenta Biotechnology, Inc.) in Research

Triangle Park, North Carolina. At Syngenta, she was involved in both research and administrative activities, serving as principal scientist, distinguished science fellow, and vice president of agricultural biotechnology. In 2002, she received one of the highest awards in her field, the Benjamin Franklin Medal in Life Sciences of the Franklin Institute. In 2013, she was also awarded the prestigious World Food Prize and, two years later, was elected to the National Inventors Hall of Fame.

Coalition for Genetic Fairness (CGF)

The Coalition for Genetic Fairness was founded in 1997 to address issues regarding the use of genetic information in insurance and employment decisions. The organization originally consisted of civil rights, disease-specific, and health care organizations, but later expanded to include representatives of industry and employers. The coalition's mission has been to educate the general public and legislators about discrimination based on genetic testing and screening. That objective was accomplished in May 2008 when President George W. Bush signed the Genetic Information Nondiscrimination Act. Since the act has become law, the coalition has refocused its efforts on informing the health community and the general public on the provisions of the act and how they can best be implemented.

Stanley N. Cohen (1935–)

Cohen and Herbert Boyer performed one of the classic studies in the early history of DNA technology in the early 1970s when they found a way to insert a foreign gene into a plasmid, a double-stranded circular piece of DNA found in bacteria and protozoa. This technology allowed Cohen and Boyer to make duplicate copies (clones) of precise segments of DNA from any given source.

Stanley Norman Cohen was born in Perth Amboy, New Jersey, on February 17, 1935. He attended Rutgers University, from which he received his BA in biological sciences in 1956, and the University of Pennsylvania School of Medicine, from which he received his MD in 1960. He then did his residency, internship, and medical research at a number of institutions, including University Hospital at the University of Michigan, Mt. Sinai Hospital in New York City, Duke University Hospital, and the National Institute of Arthritis and Metabolic Diseases. In 1968, Cohen was appointed assistant professor of medicine at Stanford University, where he spent the remainder of his academic career. He is currently professor of genetics and professor of medicine at the Stanford School of Medicine.

Cohen and Boyer met at a conference in Hawaii on bacterial plasma and over lunch discovered that their research interests melded with each other beautifully. At the time, Cohen was studying methods for inserting plasmids into bacteria in order to study their ability to develop resistance to certain antibiotics. Boyer was working on the development of certain kinds of enzymes that cut DNA into precisely defined segments with "sticky" ends. The two combined their skills to develop a method for inserting precise DNA segments into plasmids and then inserting those plasmids into bacteria. The modified bacteria could, by this method, be "engineered" to produce any desired protein product specified by the inserted DNA and, in this regard, marked the beginning of industrial biotechnology based on rDNA molecules.

In his long and illustrious career, Cohen has received a number of awards and honors, including the National Medal of Science, the Lemelson-MIT Prize, the National Biotechnology Award, and the Albert Lasker Basic Medical Research Award.

Council for Responsible Genetics (CRG)

The Council for Responsible Genetics is a nonprofit organization founded in Cambridge, Massachusetts, in 1983 for the purpose of providing a forum for discussions on the social, ethical, and environmental implications of genetic technologies. The council's activities are currently focused on eight major topics: gene patents; cloning and human genetic manipulation; women and biotechnology; genetic determinism, genetic testing, privacy, and discrimination; biowarfare and biolab safety; other genetic issues; and biotechnology—animals and agriculture. The council has produced a very large number of books, reports, brochures, position papers, online articles, and other resources on the whole range of topics in which it is interested. One of the most valuable of its publications is the bimonthly magazine *GeneWatch*, which is dedicated to monitoring the social, ethical, and environmental consequences of biotechnology.

Francis Crick (1916–2004)

Crick received a share of the 1962 Nobel Prize for Physiology or Medicine for his discovery, with James Watson, of the molecular structure of the DNA molecule. This research marked a turning point in the history of genetics, since it clarified for the first time the precise nature of a gene. Having long been thought of as some ambiguous "black box" by which hereditary characteristics are passed from one generation to the next, the gene was now describable as a discrete chemical molecule whose structure and behavior could be analyzed, understood, and, at least in theory, manipulated. Almost all forms of DNA technology would be impossible without this information.

Francis Harry Compton Crick was born in Weston Favell, England, on June 8, 1916. He received his BS in physics at University College, London, in 1937 and then began a PhD program in physics under the eminent E. N. da C. Andrade. That work was interrupted by World War II, during which time Crick served as a researcher for the British Admiralty on magnetic and acoustic mines. In 1947, he left the Admiralty and returned to his studies, but this time in the field of biology. Like many physical scientists of the time, Crick had become fascinated by the puzzles posed by biological phenomena and the increasing power of chemical and physical technologies for solving these problems. He eventually was assigned to the Cavendish Laboratory at the University of Cambridge, where he came under the influence of Sir Lawrence Bragg, one of the founders of the science of X-ray crystallography. X-ray crystallography is a technique for using X-rays to elucidate the structure of complex molecules, like proteins and nucleic acids. In 1954 he received his PhD for the analysis of polypeptides and proteins with X-rays.

In 1951, Crick met James Watson, a postdoctoral student from the United States, at Cavendish. The two men found that they had many common interests and decided to collaborate on efforts to determine the molecular structure of DNA. They succeeded in that effort in late 1952 and published the results of their work in the following year.

In the years following the discovery of DNA structure, Crick turned his attention to a number of other topics, including the process by which proteins are synthesized using the code stored in DNA molecules, the nature of the DNA code itself, the origins of life, and the chemical structure and properties of the nervous system and the brain. In 1976, Crick left Cambridge to take a position at the Salk Institute for Biological Studies in La Jolla, California. Later, he was also appointed to the faculty at the University of California at San Diego. He received many awards and prizes in addition to the Nobel Prize and was knighted by Queen Elizabeth II in 2002. Crick died in San Diego of colon cancer on July 28, 2004.

Karl Deisseroth (1971–)

In 2004, Stanford psychiatrist and researcher Karl Deisseroth and two graduate students, Edward Boyden and Feng Zhang, devised the first functioning optogenetic system in a human neuron. Optogenetics is the field of science that involves the stimulation of neurons by pulses of light so as to better understand the functioning and malfunctioning of neural systems that may be responsible for a variety of mental disorders.

Deisseroth's team achieved this breakthrough by inserting the gene for a protein called channelrhodopsin-2 into mammalian neurons by traditional methods of DNA transfection. That protein occurs naturally in some types of bacteria and algae that produces those organism's natural ability to respond to flashes of light. The experiment was proof of concept for an idea originally proposed in 1979 by Sir Francis Crick, an idea that had intrigued researchers ever since, but had only slowly come to fruition through the work of Deisseroth, Boyden, and Zhang (the latter of whom are now at the Massachusetts Institute of Technology).

Karl Deisseroth was born on November 18, 1971, in Boston, Massachusetts. He attended Harvard University, from which he received his AB in biochemical sciences in 1992. He then matriculated at Stanford University, where he completed his PhD in neuroscience in 1998 and his PhD in 2000. He completed his postdoctoral studies at Cold Spring Harbor Laboratory and his psychiatry residency at Stanford Hospital and Clinics between 2000 and 2003. In 2004 he was appointed principal investigator and clinical educator in the Department of Psychiatry at the Stanford School of Medicine, during which period he completed his studies on optogenetics. In 2005, he was appointed assistant professor of bioengineering and psychiatry at Stanford, where he was later promoted to associate professor in 2009. From 2009 to 2012 he worked as Howard Hughes Medical Institute (HHMI) Early Career Investigator before returning to Stanford as D. H. Chen Professor and Chair of Bioengineering and professor of psychiatry, positions he continues to hold. In 2014, he also accepted an appointment as investigator at HHMI. From 2013 to the present, Deisseroth has also been foreign adjunct professor at the Karolinska Institutet in Sweden and, from 2015 to the present, visiting professor at Keio University in Tokyo.

Deisseroth has received a number of important honors and awards, including the Klingenstein Fellowship Award and Robert H. Ebert Clinical Scholar Award (2005), Whitehall

Foundation Award (2005), McKnight Foundation Technological Innovations in Neuroscience Award (2005), Coulter
Foundation Early Career Translational Research Award in Bio-
medical Engineering (2005), McKnight Foundation Scholar
Award (2007), William M. Keck Foundation Medical Re-
search Award (2008), Lawrence C. Katz Prize, Duke University
(2008), Schuetze Prize, Columbia University (2008), Society
for Neuroscience YIA Award (2009), Koetser Prize laureate,
Zurich (2010), Nakasone Prize laureate, International Human
Frontier Science Program/HFSP (2010), Alden Spencer Prize,
Columbia (2011), Record Prize, Baylor University (2012),
Zuelch Prize, Max-Planck Society (2012), Richard Lounsbery
Prize from the National Academy of Sciences (2013), Dickson
Prize in Science (2014), Keio Medical Science Prize (2014), Al-
bany Prize in Medicine and Biomedical Research (2015), and
Lurie Prize in Biomedical Sciences (2015).

Jennifer Doudna (1964–)

Doudna is best known, along with Emmanuelle Charpentier,
as the co-founder of the CRISPR-Cas9 for gene editing. The
procedure is the most recent and, thus far, the most effective
means of altering the structure of a single gene in a relatively
simply way. The discovery holds enormous promise for a whole
new field of DNA technology.

Jennifer Anne Doudna was born on February 19, 1964, in
Washington, D.C. Her parents were both teachers, her father
of literature and her mother of history. She moved with them
to Hawaii in 1971, where she decided in high school that she
wanted to have a career in scientific research. She attended
Pomona College, in California, from which she received her
bachelor's degree in chemistry in 1985, and Harvard University, where she earned her PhD in 1989. At Harvard, she
worked in the laboratory of Jack Szostak, who was to win the
2009 Nobel Prize in Physiology or Medicine for his research on
the structure and function of chromosomal telomeres.

After graduation, Doudna signed on for her postgraduate studies with Thomas Cech, another Nobel Laureate (1989 in Chemistry) at the University of Colorado. There she began a long and complex study of the three-dimensional structure of ribosomes, a project that she continued after accepting a position at Yale University in 1994. In 2000, she was promoted to Henry Ford II Professor of Molecular Biophysics and Biochemistry at Yale. Two years later, Doudna was named Professor of Biochemistry and Molecular Biology at the University of California at Berkeley, where she continues to be employed. Since 1997, she has also worked as a researcher at the Howard Hughes Medical institute. She remains at California, where her laboratory focuses on the study of molecular mechanisms of RNA-mediated gene regulation.

Doudna's work has been recognized with a number of important honors and awards, many in association with her CRISPR collaborator, Charpentier, including the Alan T. Waterman Award (2000), Mildred Cohn Award in Biological Chemistry of the American Society for Biochemistry and Molecular Biology (2013), Lurie Prize in Biomedical Sciences of the Foundation for the National Institutes of Health (2014), Paul Janssen Award for Biomedical Research (2014), Breakthrough Prize in Life Sciences (2014), Princess of Asturias Award (2015), Gruber Prize in Genetics (2015), and Canada Gairdner International Award (2016).

European NGO Network on Genetic Engineering (GENET)

The European NGO Network on Genetic Engineering is an international nonprofit organization chartered under Swiss law. It was founded in 1995 for the purpose of providing information on genetically modified organisms to member organizations and individual citizens and to support activities and campaigns against genetic engineering. As of 2016, 51 organizations in

27 countries were members of GENET. The organization focuses on the distribution of information on plant and animal breeding, human health, agriculture, animal welfare, and food production as they relate to biological diversity, human genetics and medicine, the environment, and socioeconomic development. GENET sponsors two meetings annually, one an annual conference and the second a member meeting or public conference on some specific topic. The primary mechanism by which GENET achieves its objectives is its online newsletter GENET-news, which provides access to journal articles, press releases, Internet documents, and member messages on a variety of topics such as: production and use of recombinant proteins; breeding and use of transgenic plants, their ecological and biodiversity risks; production and use of food made from or produced by GMOs; production and use of enzymes and other food ingredients synthesized by GMOs; patenting of genes and transgenic organisms; biopiracy; breeding and use of transgenic animals in agriculture and "gene pharming"; xenotransplantation; national and international laws covering genetic engineering; economics and company news; and human genome projects, human cloning, and human gene therapy.

Federal Bureau of Investigation (FBI)

The Federal Bureau of Investigation is the primary federal agency responsible for investigating criminal activities in the United States. Its charge extends to all crimes not otherwise assigned to some other federal agency. The agency's CODIS Unit manages the Combined DNA Index System (CODIS) and the National DNA Index System (NDIS), the nation's largest collection of DNA specimens. The unit is also responsible for facilitating the interchange of DNA information and procedures among federal, state, and local crime laboratories and certain international law enforcement crime laboratories. As of mid-2016, NDIS contained DNA records for 12,205,768 offender profiles, which included 2,258,693 arrestee profiles and 684,519 forensic profiles. CODIS had produced over 322,011 hits assisting in more than 309,614 investigations.

Food and Agriculture Organization (FAO)

The Food and Agriculture Organization is an agency associated with the United Nations. Its mission is to help developing countries transition to modern agricultural, aquacultural, and fisheries practices and to promote good nutrition for all people of all nations. The organization's work is carried out through a number of departments, including Agriculture and Consumer Protection; Economic and Social Development; Fisheries and Aquaculture; Forestry; Knowledge and Communication; Natural Resources Management and Environment; Technical Cooperation; and Human, Financial and Physical Resources. The agency's work on issues related to biotechnology is generally not limited to a specific department, but is included as part of the work carried out within these general rubrics.

Rosalind Franklin (1920–1958)

Franklin was a biophysicist and X-ray crystallographer whose images of the DNA molecule provided invaluable information used by James Watson and Francis Crick in their elucidation of the structure of the molecule.

Rosalind Elsie Franklin was born at Notting Hill, London, on July 25, 1920, into a prominent and politically active Jewish family. Her father was Ellis Arthur Franklin, a successful banker who taught physics and history at the Working Men's College during the evening. After graduating from St. Paul's Girls School in London in 1938, Franklin enrolled at Newnham College at the University of Cambridge. Three years later, she completed her course of study, but was granted only a titular degree since women were not granted the same BA as were men. She continued her studies at Cambridge and in 1941 was granted her PhD in chemistry.

During World War II, Franklin worked at the British Coal Utilisation Research Association, studying the properties of coal, a topic on which she eventually became an internationally recognized authority. When the war ended, Franklin accepted an appointment at the Laboratoire central des services chimiques de l'État in Paris, where she was first introduced to the use of X-ray crystallography for the analysis of complex structures. She remained in France until 1950, when she was offered a job as a research associate at King's College, London, under the famous British biophysicist John Randall. Her assignment there was to study the crystalline structure of DNA fibers.

Franklin's tenure at King's produced the most remarkable images of DNA yet produced, but her work there was suffused with controversy and confusion. By late 1952 she had, in fact, obtained images that made very clear the true structure of the DNA molecule, a fact that she was not quite willing to accept, but that James Watson and Francis Crick quickly understood.

The shabby way in which Franklin was treated by her male colleagues at King's was never fully understood until the publication of Watson's story of the account of the DNA discovery in his book *The Double Helix*, published in 1968. By that time, Franklin had been dead for a decade, having died on April 16, 1958, in Chelsea, London, of ovarian cancer. Today she is memorialized by the Rosalind Franklin University of Medicine and Science, formerly the Chicago Hospital College of Medicine. Among the more than two dozen honors, awards, and other forms of recognition Franklin, the most recent is the addition of the Rosalind Franklin Lecture by the British Humanist Association to its list of annual lectures in 2016.

Genetic Alliance

Genetic Alliance is a 501(c)(3) nonprofit organization that seeks to promote the use of new genetic technologies in the treatment of human diseases. It is a collaborative of community, governmental, industrial, and private agencies with an interest in this issue. The three major programs through which Genetic Alliance conducts its work are Biotrust, Expecting Health, and Genes in Life. Biotrust provides tools with which citizens can contribute data, samples, and energy to transform health systems. Expecting Health helps families make sensible decisions about health issues at any and all stages of life development. Genes in Life is designed to provide information to individuals and families about the ways in which one's genetics affects his, her, or their personal health.

Greenpeace

Greenpeace was founded in 1971 when a group of environmentalists leased a small fishing vessel to protest nuclear tests being conducted off the coast of Alaska. Since that time, the organization has grown to become one of the largest environmental groups in the world with national chapters in the United States and more than 40 other countries. The organization's mission has also expanded to include a number of environmental issues, including global climate change, the oceans, ancient forests, peace and disarmament, nuclear weapons, genetic engineering, toxic chemicals, and sustainable trade. Among the many victories claimed by the organization are: a successful effort to reduce the destruction of forests in Indonesia for the construction of palm oil plantations; passage of a ban on incandescent light bulbs in Argentina; adoption of a ban on deep-sea ocean bottom trawling off the coast of Chile; a successful public vote on a ban on genetically engineered crops in Switzerland; discontinuation of all genetically engineered crops in India by the Bayer Corporation; and suspension of tests on genetically engineered Roundup by the Monsanto corporation.

Woo-suk Hwang (1953–)

For a period of time in 2004 and 2005, Hwang was widely regarded as one of the most successful and influential cloning researchers in the world. In a pair of papers published in the journal *Science* in those two years, he reported that he had been successful in cloning human embryos, from which he was then able to extract human embryonic stem cells. Hwang's research was praised widely by his colleagues and peers because it opened a new door to the cloning of human embryos for therapeutic purposes. He received wild praise from both the South Korean government and the general populace for these supposed discoveries. Only a year after his second paper appeared, however, Hwang was charged with falsifying his research results and charged with embezzlement of government funds and violations of the nation's bioethics laws. In 2007, he was fired from his position at Seoul National University and was banned from conducting further stem cell research and from receiving federal funding for related research.

Woo-suk Hwang was born on January 29, 1953, in the village of Bu-yeo in the province of Chungnam, South Korea. His father died when he was five years old, and he began to work on the family farm to help support his widowed mother and five siblings and to earn enough money to continue his own studies. After graduating from Daejeon High School, he matriculated at Seoul National University, from which he received his bachelor's degree in veterinary medicine in 1977 and his master's degree and doctorate in theriogenology in 1979 and 1982, respectively. (Theriogenology is the study of animal reproduction.) For his postdoctoral studies, Hwang traveled to Hokkaido University in Sapporo, Japan, as a visiting fellow. In 1986, he accepted an offer to join the faculty at Seoul National University, where he remained until being expelled in 2007.

During his years at the National University, Hwang specialized in the reproduction of cattle and became especially interested in artificial reproductive technologies (ARTs), such as in vitro fertilization (IVF) and cloning. In 1993, he announced the birth of the first calf produced by IVF, and, later, the first cloned cow (1999), the first cloned pig (2002), and the first cow resistant to so-called mad-cow disease (2003). This research culminated, of course, in his reported (but later refuted) claim of cloning the first human embryo.

The story of Hwang's disgrace appeared to represent the end of his academic and research career. But such has hardly been the case. Even before leaving his post at National University, Hwang had become involved in the creation of a new and private endeavor, the Sooam Biotech Research Foundation, located in Seoul. The purpose of the foundation was to support and carry out research on cloning of a number of animals for a variety of purposes. Over the next decade, Sooam and Hwang have reported a number of significant accomplishments in the field, including the cloning of the first companion dog, Missy, in October 2007; the commercial cloning of the first companion dog, Lancelot, in November 2008; the cloning of dog disease models for diabetes and Alzheimer's disease in 2010; the initiation of a program to clone the extinct wooly mammoth in 2012; and the cloning of a Chinese champion Tibetan mastiff in 2014. By 2016 Sooam Biotech was regularly cloning domestic pigs with genetic predisposition for certain diseases, animals that could, therefore, be used in research on those diseases. The company's most significant commercial accomplishment thus far may have been an arrangement with the Chinese company Boyalife for the construction of a factory capable of producing 100,000 cloned cattle per year.

Innocence Project

The Innocence Project was established in 1992 at the Benjamin N. Cardozo School of Law of Yeshiva University in New York City by Barry C. Scheck and Peter J. Neufeld. Both men had been involved in cases involving the use of DNA typing from its earliest days. The purpose of the project was to assist individuals who had been improperly convicted of crimes. As of early 2016, 337 individuals in the United States had been exonerated of the crimes for which they had been convicted, based on DNA evidence, and an additional 300 cases were currently being investigated. In addition to working with individual prisoners who claim they have been wrongfully convicted, the project works for the passage of enlightened legislation at the state and federal levels to make possible the exoneration of innocent victims of the law.

Institute for Responsible Technology (IRT)

The Institute for Responsible Technology was founded in 2003 by Jeffrey Smith, a former marketing consultant who became interested in the risks posed by genetically modified foods in the human diet. The mission of IRT is to educate the general public about the risks of eating GM foods and about healthier alternatives to engineered foods. IRT's efforts to exclude GM foods from the American diet are expressed in its Campaign for Healthier Eating in America, whose goal is to stop the use of genetic engineering in the American food industry. The institute is also involved in the development of a GM-free schools movement, which focuses its non-GMO foods efforts on the foods served in American schools. IRT's Web site is a valuable resource for information on the fundamentals of genetically modified organisms and foods, the use of recombinant bovine growth hormone (rBGH) in dairy products, ways of shopping for non-GMO foods, and opportunities for activism against engineered foods.

International Bioethics Committee (IBC)

The International Bioethics Committee (IBC) was created in 1993 to provide a forum for the study of developments in the life sciences and their potential implications for human dignity and freedom. The committee consists of 36 experts in the field who are appointed by the director general of the United Nations Educational, Scientific and Cultural Organization (UNESCO) for a renewable term of four years. They come from a number of different fields, including medicine, genetics, chemistry, law, anthropology, philosophy, and history. The committee has drafted reports on a wide variety of topics ranging from intellectual property rights to plant biotechnology to genetic transformations in plants.

International Biopharmaceutical Association (IBPA)

The International Biopharmaceutical Association was founded in 1990 to bring together institutions and corporations from nations around the world to work on issues of common concern. The association collaborates with international, regional, and national bodies concerned with the biopharmaceutical industry and clinical research. Its services are available to organizations, institutions, regulatory authorities, individual policy and decision-makers, specialists, administrators, researchers, educators, and students. As of 2016, IBPA had more than 7,000 individual and 200 corporate members from every part of the world. The association takes part in a number of meetings, conferences, and events at companies, academic institutions, and other institutions every year. A major part of IBPA's focus is career counseling. It provides information on careers in biopharmaceuticals, lists job openings, offers assistance with resume writing and job searching, provides coaching for careers, and provides information on internships.

International Centre for Genetic Engineering and Biotechnology (ICGEB)

The International Centre for Genetic Engineering and Biotechnology was established in 1983 as the result of a treaty negotiated through the United Nations. Under the provisions of the treaty, the organization actually began operations in 1994 when the treaty was ratified by the required number of nations. At that point, the center became an autonomous intergovernmental agency. Its purpose is to conduct research in the field of genetic engineering and biotechnology that will have useful applications for developing nations, especially in the fields of biomedicine, crop improvement, environmental protection and remediation, biopharmaceuticals, and biopesticide production. The center originally had two components, one in Trieste, Italy, and one in New Delhi, India. In 2007, a third facility at Cape Town, South Africa, was added. As of early 2016, 60 nations were members of the center, with an additional 17 nations awaiting ratification of the founding treaty. The center employs about 400 people from 38 different countries as research scientists, postdoctoral fellows, doctoral students, research technicians, and administrative personnel. It has produced more than 2,700 publications in peer-reviewed scientific journals around the world.

Sir Alec Jeffreys (1950–)

Jeffreys is discoverer of the restriction fragment length polymorphism (RFLP) technique for the amplification of small segments of DNA, a key technology in a number of fields of DNA technology, including forensic science, the determination of paternity, and wildlife biology. RFLP has been replaced in many instances by the polymerase chain reaction (PCR) procedure developed by Kary Mullis, but it was the first such technology to be used and is still employed in some circumstances.

Alec John Jeffreys was born on January 9, 1950, in Oxford, England. He was very interested in science as a child and was given a number of pieces of scientific apparatus with which to experiment while he was still in grammar school. He enrolled at Merton College, Oxford University, from which he received his BA in biochemistry in 1972 and his PhD in genetics in 1975. After completing his postdoctoral studies at the University of Amsterdam, Jeffreys accepted an appointment in the Department of Genetics at Leicester University, where he has remained ever since. Jeffreys sometimes pinpoints the exact moment at which he understood the potential value of RFLP technology: 9:05 a.m. on September 10, 1984. He was examining an X-ray "fingerprint" of a lab technician when he realized that such fingerprints can be used to precisely identify a person's genetic identity. He then derived the details by which such a procedure could be conducted and published the results in a paper titled "Hypervariable 'Minisatellite' Regions in Human DNA," which appeared in the journal *Nature* on March 7, 1985.

Jeffreys has continued his research on a number of topics in genetics, including eukaryotic introns, the evolution of gene families, and the process of genetic recombination. He has received a number of honors, including election to the U.S. National Academy of Sciences, the Colworth Medal for Biochemistry of the Biochemical Society, and the Gold Medal for Zoology of the Linnean Society of London. He was knighted by Queen Elizabeth II in 1994. In 2014, Jeffreys was awarded the prestigious Copley Medal of the Royal Society.

Gregor Mendel (1822–1884)

Mendel is often called the father of genetics because of his research on the transmission of physical characteristics in pea plants. Based on his experiments, Mendel elucidated a set of laws, now called Mendel's Laws of Genetic Inheritance. The first law is called the Law of Segregation, namely that each parent contributes one and only one allele to the formation of an egg or sperm. The second law, the Law of Independent Assortment, says that allele pairs in a parent separate from each other independently during reproduction, so that physical traits are also transmitted independently.

Mendel was born Johann Mendel in Heinzendorf bei Odrau, Silesia, then part of the Austrian Empire and now the city of Hynčice, Czech Republic. He worked on the family farm, where one of his chores was tending fruit trees for the lord of the local manor. Mendel attended the Philosophical Institute in Olomouc from 1840 to 1843 before entering the Augustinian Abbey of St. Thomas in Brno (Brünn). He took the name of Gregor upon becoming a monk. Mendel was ordained as a priest in 1847 and, four years later, sent to the University of Vienna to study mathematics and science in preparation for a career as a teacher in local Catholic schools. After failing his examinations at Vienna three times, he was finally assigned as a science teacher at a lower school, the Brno Realschule in 1854.

In 1857, Mendel began his historic studies of pea plants in his garden at the Brno monastery, research that continued for eight years. Over time, he grew more than 10,000 pea plants and made careful notes of the pattern of inheritance among them. His genius was to combine careful experimental skills with a skill for mathematical analysis of the results of his work. Mendel reported the results of his research at two meetings of the Natural History Society of Brno on February 8 and March 8, 1865, and then published those results in a paper, "Versuche über Pflanzen-Hybriden ("Experiments on Plant Hybridization"), in the Proceedings of the Natural History Society of Brno for the year 1865 (published in 1866). The scientific community paid almost no attention to Mendel's research, and his paper was cited only three times in the next 35 years. His work remained essentially unknown until it was re-discovered in 1900 almost simultaneously by three different researchers, Dutch botanist Hugo de Vries, Austrian botanist Erich von Tschermak, and German botanist Carl Correns.

Mendel's scientific work essentially ended in 1868 when he was elected abbot at St. Thomas. Administrative responsibilities associated with his new position, aggravated by ongoing disputes with civil authorities over taxation issues, left him little time for non-monastic issues. He died in Brno on January 6, 1884, from chronic nephritis.

Johannes Friedrich Miescher (1844–1895)

Miescher was a biochemist best known for his discovery of the first nucleic-acid-like substance, which he named *nuclein*, in 1869.

Usually known by his middle name, Friedrich Miescher was born on August 13, 1844, in Basel, Switzerland. Both his father and his uncle had held the chair of anatomy at the University of Basel, and his father encouraged Friedrich to pursue a medical degree. Friedrich decided, however, that his partial deafness, resulting from a case of typhus, might preclude a career in medicine. When he earned his MD from Basel in 1868, then, he decided to focus on biochemical research rather than medical practice. After graduation, he took a position in the laboratory of the eminent German biochemist Ernst Felix Hoppe-Seyler at the University of Tübingen, where he began research on the chemical composition of the cellular nucleus. In his studies, Miescher discovered a new chemical compound of unusual properties. It was similar in some way to proteins, with molecules of very large size, but it was rich in phosphorus and lacking in sulfur. Proteins never contain phosphorus, but always contain some sulfur. Hoppe-Seyler was so surprised by Miescher's results that he would not allow his student to publish until he, Hoppe-Seyler, had repeated and confirmed those results.

Miescher called the substance he discovered *nuclein*, a material we now know as a complex of pure deoxyribonucleic acid (DNA) and a variety of supportive proteins typically found with DNA. In later research, Miescher was able to prepare pure DNA, although he was never able to determine its precise biochemical function.

Miescher died in Davos, Switzerland, on August 16, 1895, at the age of 51 from tuberculosis.

Juan Francisco Martinez Mojica (1963–)

Of the various methods currently available for gene editing, probably the most popular at the present time is the CRISPR-Cas9 system (for "clustered regularly interspaced palindromic repeats"). Many researchers have contributed to the current methodology used in CRISPR-Cas9, but certainly one of the early pioneers was Spanish researcher Juan Francisco Martinez Mojica. In 1993, Mojica and his colleagues first described the basic structure of bases in DNA that constitute the mechanism by which prokaryotes carry out repair processes when attacked by harmful agents. He later suggested the name for these repeating units, CRISPR, a name that has been retained to this day. Mojica, now at the University of Alicante, has spent the greatest part of his academic life in learning more about the CRISPR method for editing genes.

Juan Francisco Martinez Mojica was born in Elche, on the eastern coast of Spain, on October 5, 1963. He completed his early studies at the Primary Academy Andes and the National Mixed College of Vázquez de Mella. He then continued his studies at the University of Murcia and the University of Valencia, from which he earned degrees in biology in 1986. Mojica then entered the University of Alicante for his doctoral studies, earning his PhD in biology in 1993 with a thesis on the influence of environmental factors on the DNA and gene expression in the genus Haloferax.

Mojica completed his postdoctoral studies at Alicante in 1994 and at Oxford University in the United Kingdom from 1995 to 1996, where he began his studies on the CRISPR system used by bacteria to protect themselves against invasion by viruses and bacteriophages. By 2005 he had determined the mechanism by which this process occurs, laying out the general principles for gene editing using the CRISPR-Cas9 system now widely used throughout the world.

In December 1994, Mojica was named Profesor Titular Interino de Universidad (associate professor) in the Department of Physiology, Genetics, and Microbiology at the University of Alicante, where he has remained ever since. In 1997 he was promoted to Profesor Titular de Universidad (professor) at Alicante. Mojica has also conducted research at the University of Paris (1991–1992), the University of Utah (1993), and the University of Oxford (1995–1996).

Kary Mullis (1944–)

Mullis is the inventor of a technology called the polymerase chain reaction (PCR) for the amplification of DNA segments, an accomplishment for which he was awarded the 1993 Nobel Prize in Chemistry. The technology is now a key procedure in many fields of DNA technology, because it allows the analysis of very small amounts of DNA after a series of steps that makes them large enough to study.

Kary Mullis was born in Lenoir, North Carolina, on December 24, 1944. He attended the Georgia Institute of Technology, from which he received his BS in 1966, and the University of California at Berkeley, where he earned his PhD in biochemistry in 1973. He then completed two postdoctoral programs in pediatric cardiology at the University of Kansas Medical School and in pharmaceutical chemistry at the University of California at San Francisco. In 1979, Mullis took a position with the Cetus Corporation in Emeryville, California, one of the first companies formed to develop industrial, pharmaceutical, and other applications of new DNA technology. During his seven years at Cetus, Mullis worked on and developed the PCR technology for which he is best known. After leaving Cetus in 1986, Mullis took a job as director of molecular biology at the Xytronyx corporation in San Diego, a post he held for two years. He then began consulting on nucleic acid chemistry for a number of corporations, including Abbott Labs, Angenics, Cytometrics, Eastman Kodak, Milligen/Biosearch, and Specialty Laboratories. Mullis is also known for a somewhat free-spirited view of life and some unconventional views on scientific issues of social significance, such as the causes of HIV/AIDS disease and global climate change. He has written a popular autobiography, *Dancing Naked in the Mine Field*. He has largely been inactive in research over the past two decades.

National Human Genome Research Institute (NHGRI)

The National Human Genome Research Institute was founded in 1989 as the National Center for Human Genome Research.

Originally its function was to carry out the responsibilities of the National Institutes of Health in the Human Genome Project. As that project evolved, the mission of NHGRI changed, expanding in 1993, for example, to include a study of genetic diseases. In 1996, its mission was expanded further with the creation of the Center for Inherited Disease Research (CIDR) to study the genetic component of complex inherited disorders. In 1997, the center was given its present name and was elevated to a full institute within the National Institutes of Health (NIH), putting it on equal basis with 26 other institutes and centers that make up the NIH. Upon completion of its first mission, sequencing of the complete human genome in April 2003, the institute has moved on to other goals and objectives, including research on the ethical, legal, and social implications (ELSI) of genomic discoveries; developing strategies for using genomic information to reduce health disparities among Americans; and providing support for the training and education of the next generation of genomic researchers. Among the topics of current interest are coverage and reimbursement in genetic tests; human subjects research; genetic discrimination; regulation of genetic tests; privacy in genomics; informed consent; intellectual property and genomics; genetics and public policy; and genome statute and legislation database.

National Society of Genetics Counselors (NSGC)

The National Society of Genetics Counselors was founded in 1979 to be the primary advocacy group for genetics counselors and to ensure the availability of quality genetic counseling through education, research, and the promotion of public policy in the field. Important elements of the organization's work are carried out through an annual educational conference, occasional regional conferences, an online educational conference, and continuing education courses. The organization's Web site provides information and resources for health care professionals and for the general public. It also explains the process by which one becomes a genetics counselor, with referrals to programs available in the field.

Marshall Nirenberg (1927–2010)

Nirenberg was awarded a share of the 1968 Nobel Prize in Physiology or Medicine for his discovery of the first codon, a set of three bases that codes for a specific amino acid. Earlier research had showed that the nitrogen bases on a DNA molecule acting as a triplet direct the synthesis of specific amino acids. That is, the combination CGA (cytosine-guanine-adenine) somehow "tells" a cell to make a specific amino acid. The problem was that no one yet knew which codons coded for which amino acids.

Working with Heinrich J. Matthaei at the NIH, Nirenberg obtained the answer to that question. The Nirenberg-Matthaei experiment was conceptually relatively simple. They prepared RNA molecules consisting only of the nitrogen base uracil (UUU). They then inserted those molecules into a sample of *E. coli* bacteria that was treated to destroy their natural DNA, ensuring that the only compounds synthesized were those coded for by the UUU triplet. Upon analysis, Nirenberg and Matthaei found that the only compound produced by the bacterial cells was the amino acid phenylalanine. It was evident, then, that the triplet UUU codes for the amino acid phenylalanine. The first element in the genetic code had been determined.

Marshall Warren Nirenberg was born in New York City on April 10, 1927. When he developed rheumatic fever, the family moved to Florida, where Marshall soon developed an intense interest in the natural world. He began to collect an extensive set of notes and diaries about his own research on living organisms. In 1945, Nirenberg enrolled at the University of Florida at Gainesville, from which he received his BS in chemistry and zoology in 1948 and his MS in zoology in 1952. Nirenberg then continued his studies at the University of Michigan in

Ann Arbor, which granted him his PhD in biological chemistry in 1957. He then completed two postdoctoral programs at the National Institute of Arthritis, Metabolism, and Digestive Diseases (NIAMDD). It was at NIAMDD that Nirenberg carried out his work on the genetic code with Matthaei.

In 1962, Nirenberg was offered the position of chief of the section of biochemical genetics at the National Heart Institute (now the National Heart, Lung, and Blood Institute; NHLBI) of the NIH. He remained with the NHLBI for the remainder of his academic career. In addition to the Nobel Prize, Nirenberg was awarded the National Medal of Science, the National Medal of Honor, and the Molecular Biology Award of the National Academy of Sciences. In 2001 he was elected to the American Philosophical Society. Nirenberg died in New York City on January 15, 2010, after a short battle with cancer.

Non-GMO Project

The Non-GMO Project originated at a single grocery store, The Natural Grocery Company, in Berkeley, California, in 2003. It was created because of concerns by customers that they were uncertain as to the presence of genetically modified foods at Natural Grocery. The owners of Natural Grocery decided that they needed clear standards for defining genetically modified foods and an independent, disinterested source for verifying the status of GM and non-GM foods. They were eventually joined in this effort by The Big Carrot Natural Food Market in Toronto, Ontario, which had been working for well over a year to develop an identification and labeling process. In 2007, Non-GMO had expanded to include all stakeholder groups in the natural products industry, including consumers, retailers, farmers, and manufacturers. The organization also began to form a number of advisory boards on a number of technical and policy issues. The process by which food products can be verified as non-GMO and listed as approved by the Non-GMO

Project, as well as a list of participants in the program, is available on the organization's Web site.

Northwest Resistance against Genetic Engineering (NW RAGE)

Northwest Resistance against Genetic Engineering describes itself as "a non-violent, grassroots organization dedicated to promoting the responsible, sustainable and just use of agriculture and science . . . [that is] working towards a ban on genetic engineering and patents on life." Some issues on which NW RAGE is currently working are agrofuels, genetically engineered trees, recombinant bovine growth hormone (rBGH), Roundup Ready creeping bent grass, and safe foods. Some specific goals of the organization are a ban on genetic engineering, a ban on patents on any kind of life form, a ban on biopiracy (defined as the theft of indigenous people's genes and knowledge), a ban on cloning of humans and animals, a rescinding of all current FDA approvals for genetically engineered products on the market, and the cessation of factory farming. The organization's Web site is a rich resource of shopping resources, industrial contacts, films and podcasts, downloads, and local resources. It also provides access to hundreds of articles dealing with the human health and environmental aspects of genetic engineering.

Office of Biotechnology Activities (OBA)

The Office of Biotechnology Activities is an agency within the NIH responsible for promoting science, safety, and ethics in the development of public policies in three specific areas: biosafety, biosecurity, and biomedical technology assessment. Biosafety refers to a host of activities carried out to ensure that research funded by the NIH in the United States and overseas is conducted under the highest standards of safety for researchers, the general public, and the environment. Biosecurity activities are designed to ensure that biotechnology research is not used for purposes that will threaten the health or safety of the general public. The biomedical technology assessment arm of the agency's work is directed at assessing the scientific, ethical, and social implications of biotechnology research, including the incorporation of such work into scientific and clinical practice.

Office of Public Health Genomics (OPHG)

The Office of Public Health Genomics is a division of the Centers for Disease Control and Prevention (CDC) established in 1997 as the Office of Genetics and Disease Prevention, later renamed the Office of Genomics and Disease Prevention in 2003, and then given its present name in 2006. The agency's mission is to integrate the new field of genomics into public health research, policy, and programs in such a way as to improve the lives and health of all people. A topic of recent interest to OPGH has been the use of genetic testing in the diagnosis of diseases in newborn children, adults, and human fetuses. Some topics of recent special interest to OPHG have been Down syndrome, tuberculosis, food safety, bleeding disorders, pathogen genomics, genomic testing, and epidemiology. The office maintains an excellent database of information on many aspects of health genomics.

Organic Consumers Association (OCA)

The Organic Consumers Association was formed in 1998 in reaction to proposed regulations for organic foods proposed by the USDA. The organization has continued to work for the creation and enforcement of strict standards for the growing and labeling of organic foods. It claims to be the sole national organization working on behalf of "the nation's estimated 50 million organic and socially responsible consumers." OCA's political efforts are guided by a six-point platform adopted in 2005 that outlines the organization's goals and objectives. OCA maintains a news section page on its Web site that contains print and electronic articles on topics related to genetically engineered foods.

Ingo Potrykus (1933–)

In collaboration with Peter Beyer, Potrykus invented golden rice, a genetically engineered agricultural product capable of delivering increased amounts of vitamin A to people who eat it. The product potentially has very significant benefits since as many as 250,000 to 500,000 children around the world suffer from blindness and death because of suboptimal levels of vitamin A in their daily diet.

Ingo Potrykus was born in Hirschberg, Germany, on December 5, 1933. He attended the universities of Cologne and Erlangen, where he studied botany, biochemistry, zoology, genetics, philosophy, and physical education. He received his PhD in plant genetics from the Max-Planck-Institute for Plant Breeding Research in Cologne in 1968. He was then offered a position at the Institute of Plant Physiology in Stuttgart-Hohenheim, where he remained from 1970 to 1974. Potrykus then took a position as research group leader at the Max-Planck-Institute for Genetics at Ladenburg-Heidelberg from 1974 to 1976 before assuming a similar position at the Friedrich Miescher Institute in Basel, Switzerland. In 1986, he left the Friedrich Miescher Institute to become professor of plant sciences at the Eidgenössische Technische Hochschule Zürich (Swiss Federal Institute of Technology; ETH), where he remained until his retirement in 1999.

Throughout his career, Potrykus has been interested in finding ways of using genetic engineering to increase food production for the world's growing population. He was especially interested in solving the problem of adding vitamin A to the rice grains on which so much of the world's population depends for its primary food. Rice plants do synthesize the precursors of vitamin A in their vegetative tissues, but not in rice grains themselves. Potrykus and his colleagues found that the addition of genes for two additional enzymes to the rice plant, phytoene synthase and phytoene desaturase, made possible the production of vitamin A in rice grains as well as in the plant's tissues.

Solving the scientific problem of producing golden rice was only the beginning for Potrykus and his followers. Since laboratory success was achieved, the Golden Rice project has been attempting to complete other requirements needed before the product can actually be used in the field, such as gaining necessary financial support and licenses and conducting field and nutrition tests. In 2001, Potrytus transferred all rights to Golden Rice to the Syngenta corporation, providing a commercial platform through which the product could be distributed to those who need it. As of March 2016, however, Golden Rice had yet to be made available to any users anywhere in the world, largely as the result of objections by some environmentalists, potential competitors, and others with objections about the product.

Presidential Commission for the Study of Bioethical Issues

The Presidential Commission for the Study of Bioethical Issues was created by executive order of President George W. Bush and was originally known as the Council on Bioethics. The council was renewed by new executive orders by President Bush in 2003, 2005, and 2007. The current Commission was created by executive order of President Barack Obama in November 2009. It is the latest of a number of federal commissions appointed to consider bioethical issues, dating back to 1974, with the creation of the National Commission for the Protection of Human Subjects of Biomedical and Behavioral Research by the U.S. Congress. A history of earlier commissions and reports they produced is available at the council's Web site at http://bioethics.gov/history. In its history, the commission has considered a number of bioethical issues related to topics such as aging and end of life, biotechnology and public policy, cloning, death, genetics, health care, human dignity, nanotechnology, neuroethics, newborn screening, organ transplantation, research ethics, sex selection, and stem cells.

Hamilton O. Smith (1931–)

Smith was awarded a share of the 1978 Nobel Prize for Physiology or Medicine for his research on restriction enzymes. Restriction enzymes are chemical compounds that recognize specific nucleotide sequences in a nucleic acid and then cleave the molecule at those sites. More than 3,000 restriction enzymes are currently known and about 600 are commercially available. Restriction enzymes are essential in most DNA technologies, because they allow a researcher to cut DNA and RNA molecules in precise locations, permitting the insertion of DNA or RNA segments from other sources in the production of transgenic organisms.

In his research, Smith and his colleagues identified the first restriction enzyme, found in the bacterium *Haemophilus influenzae*, which they called endonuclease R, but which is now known as HindII. Shortly thereafter, they also discovered the DNA sequence that the enzyme recognizes and cuts:

GT(T/C)(A/G)AC
CA(A/G)(T/C)TG

Hamilton Othanel Smith was born in New York City on August 23, 1931. In 1937, his father joined the faculty of the Department of Education at the University of Illinois (UI), and Hamilton became a student at the University Laboratory School at UI. After he graduated from high school, Smith enrolled at the UI, planning to major in mathematics, for which, he later wrote, "I had a flair but no deep talent." In 1950, he transferred to the University of California at Berkeley (UCB), where, although he continued his studies in mathematics, he discovered that his real passion was biology and biochemistry. After receiving his degree from UCB in 1952, Smith applied to the Johns Hopkins University Medical School and was accepted. He was awarded his MD there in 1956.

After completing his internship at Barnes Hospital in St. Louis, his residency at University Hospital at the University of Michigan, and a tour of duty as researcher in human genetics at Michigan, Smith accepted an appointment as assistant professor of microbiology at Johns Hopkins University. He remained at Johns Hopkins until his retirement in 1998, when he was named American Cancer Society Distinguished Research Professor Emeritus of Molecular Biology and Genetics. In 1994, Smith collaborated with J. Craig Venter at The Institute for Genomic Research to sequence the complete genome of the *H. influenzae* bacterium. In July 1998, Smith joined a research team at Celera Genomics, where he worked on sequencing the human genome and the genome of the common fruit fly. In November 2002, he left Celera to join the J. Craig Venter Institute, where he is leader of the synthetic biology and bioenergy research groups. In March 2016, Smith and Venter collaborated on the creation of a new bacterial life form with a smaller genetic code than any known in the natural world, presumably the minimal code needed for life.

Robert A. Swanson (1947–1999)

Swanson was a venture capitalist who founded the biotechnology firm Genentech in 1976 with Herbert W. Boyer. He was affiliated with the company for almost the rest of his life, retiring in December 1996.

Robert Arthur Swanson was born on November 29, 1947, in Brooklyn, New York. He attended the Massachusetts Institute of Technology (MIT), from which he received his SB in chemistry and SM in business management, both in 1970. He then took a job with Citicorp Venture Capital Limited, where he remained until 1974. He then joined the venture capital firm of Kleiner, Perkins, Caufield & Byers (KPDB) in Menlo Park, California. A year later, aided by $100,000 in seed money from KPDB, Swanson and Boyer founded Genentech for the purpose of developing new recombinant DNA products and developing them for the marketplace. He served as chief executive officer at Genentech from 1976 to 1990 and as chairman of the board from 1990 to 1996. Swanson then left Genentech to continue venture capital activities on his own, eventually becoming involved in the formation of a new biotechnology company, Tularik, Inc. In 1998, Swanson fell ill with brain cancer, a disease from which he died on December 6, 1999, at his home in Hillsborough, California.

Dizhou Tong (1902–1979)

For many people, including most of the population of China who know about cloning, Tong is legitimately entitled to be called the father of cloning. During the mid-20th century, he cloned a number of organisms, including, in 1963, an Asian carp, at the time, the most complex organisms to have been created by the process of somatic cell nuclear transplantation (SCNT). Tong achieved this step more than three decades before researchers at the Roslin Institute in Scotland cloned the first mammal, the sheep Dolly, for which many historians feel justified in calling Tong the father of cloning.

Tong Dizhou (according to Chinese naming custom) was born in a small village in Jing County in Zhejiang Province, People's Republic of China, on May 28, 1902. (He is also known as Ti Chou Tung.) Tong's father was a teacher in his home village, and took responsibility for his own son's education at home until he reached the age of 14, when he died and left family responsibilities to his wife. Two years later, Tong enrolled at the Xiaoshi Middle school, where he was then the oldest pupil. After graduating from Xiaoshi in 1923, Tong enrolled at Fudan University in Shanghai, from which he received his degree in biology in 1927.

Upon his graduation from Fudan, Tong followed one of his former teachers, Bao Cai, to his first academic post as assistant in the department of biology at the National Central University in Nanjing. He remained there until 1930, when he decided to continue his studies at the Universite Libre de Bruxelles (Free University of Brussels) in Brussels, Belgium, from which he received his doctorate in biology four years later.

Tong then returned to China where he accepted an appointment as professor of biology at Shandong University in Qhingdao. The beginning of Tong's academic career at Shandong came at an less than fortuitous time only a few years before the outbreak of the disastrous (for China) Second Sino-Japanese War of 1937–1945. During the early years of that war, battles forced the university to move its operations to a number of new locations, and operations were sometimes suspended because of Japanese incursions into an area. Supplies were also difficult to obtain, and it is said that Tong sometimes did not have even the most basic equipment with which to conduct his research. The end of the war in 1945 brought more stability to China and to Tong's own career. In that year, he was promoted to dean of the department of zoology at Shandong, where, a year later, he founded the Institute of Marine Biology. The institute later became part of the Chinese Ocean University, located in Qingdao.

In 1948, the U.S. Rockefeller Foundation provided Tong with the funds necessary to spend a year in the United States, during which time he worked as a visiting scholar at Yale University and an independent investigator at the Woods Hole Marine Biological Laboratory. At the end of his sabbatical year in the United States, Tong returned to Shandong, where he was promoted to vice president of the university.

Tong returned to China at a time of renewal and revitalization culminating in the creation of the People's Republic of China in 1949. Among the many administration changes associated with the new government was the creation of the Chinese Academy of Sciences (CAS), which was to become the focus of much of Tong's career for the rest of his life. One of his first acts upon his return to Shandong was the creation of the laboratory of experimental embryology within the Institute of Experimental Biology of the new CAS. He also established a new Laboratory of Marine Biology in Qingdao, which was, within a very short period of time, to become the center of virtually all research in the field in China. Tong was chosen as the first director of the laboratory, later to become the Institute of Oceanology of the CAS, a post he held until his death in 1979. In 1955, Tong resigned his post at Shendong Shandong to become head of the Division of Biological Sciences at CAS, a post he also held for more than two decades, eventually becoming vice president of the CAS a year before his death. Throughout his long career of teaching and administration, Tong continued to carry out his research on embryogenesis, including some of the earliest and most successful experiments on somatic cell nuclear transplantation. He died on March 30, 1979.

Union of Concerned Scientists (UCS)

The Union of Concerned Scientists was formed in 1969 as a collaborative effort between students and faculty members at the Massachusetts Institute of Technology (MIT). Its earliest campaigns were against nuclear power plants and certain types of weapon systems being developed by the U.S. military. In recent years, it has greatly expanded its field of interests, now operating programs in global climate change, clean energy, clean transportation, nuclear power, nuclear weapons, food and agriculture, and invasive species. Its efforts in the area of food and agriculture involve the promotion of sustainable agriculture methods and improved government regulation of genetically engineered organisms and products. UCS lists its recent successes in this field as including the shaping of legislation supporting organic and sustainable agriculture, achieving a meaningful label for grass-fed meat, preventing a type of human medicine from being used in animal agriculture, pressing for a ban on the production of drugs and industrial chemicals in food crops, and strengthening oversight of pharma crops and cloning. The organization's Web site has nearly 800 articles on a variety of issues related to genetic engineering at http://www.ucsusa.org/search/site/genetic%20en gineering#.VwKowvkrJD8.

U.S. Environmental Protection Agency (EPA)

The U.S. Environmental Protection Agency is the primary agency for monitoring the environmental health of the United States. Among its many responsibilities is monitoring genetically engineered pesticides. Authority for this assignment rests in two important pieces of legislation, the Federal Insecticide, Fungicide, and Rodenticide Act (FIFRA), 7 U.S.C. §136, et seq., and the Federal Food, Drug, and Cosmetics Act (FFDCA), 21 U.S.C. §301, et seq. With the development of methods for genetically engineering plants with the ability to produce pesticides and related products, responsibility for investigating and licensing these products fell to the EPA. Detailed information about the EPA's work in this regard is available on the agency's Web site under its section on Plant Incorporated Products (PIPs), at http://www.epa.gov/opp00001/biopesticides/pips/index.htm.

U.S. Food and Drug Administration (FDA)

The U.S. Food and Drug Administration (FDA) dates to about 1848, when the U.S. Congress provided for the appointment of Lewis Caleb Beck to the U.S. Patent Office with the charge of carrying out chemical studies of agriculture products. By far the most important enabling legislation related to the FDA, however, was the 1906 Pure Food and Drugs Act, which, for the first time in American history, provided strict regulations about the production, transportation, processing, sale, distribution, and consumption of a wide variety of food products in the United States. The general principle that the U.S. federal government had an essential role to play in protecting the food supply of the American people was laid down in the Pure Food and Drugs Act and, in a sense, continues to act as the guiding principle for today's FDA.

The modern cabinet department, however, has expanded far beyond Lewis Beck's modest beginning a century and a half ago. Today the department has a civilian and military staff of about 16,000 and an annual budget of about $4.9 billion. The agency's mission is subdivided into eight major categories: food; drugs; medical devices; radiation-emitting products; vaccines, blood, and biologics; animal and veterinary; cosmetics; and tobacco products. The major divisions through which the FDA carries out its activities are the Office of the Commissioner, which is responsible for most administrative activities and also includes the National Center for Toxicological Research; the Office of Foods and Veterinary Medicine, which includes the Center for Food Safety and Applied Nutrition and the Center for Veterinary Medicine; the Office of Global Regulatory Operations and Policy, which is responsible for international operations; and the Office of Medical Products and Tobacco, which includes offices dealing with biologics research and evaluation, devices and radiologic health, drug evaluation and research, tobacco products, and special medical programs.

According to the 1986 Coordinated Framework for Regulation of Biotechnology, the FDA has responsibility for foods and food additives; human drugs, medical devices, and biologics; animal drugs; and plants and animals that have been genetically engineered in one way or another. In the three decades since the original assignment of responsibilities were assigned, and as DNA technology has grown and become more sophisticated, the details of FDA responsibilities for the regulation of genetically modified products have expanded and changed to some extent.

J. Craig Venter (1946–)

While a researcher at the NIH in the 1980s, Venter developed a method for mapping the human genome using chemical units known as expressed sequence tags (ESTs). He founded Celera Genomics in 1998 as a vehicle for using ESTs to elucidate the human genome, a program that eventually ran concurrently for many years with the Human Genome Project.

John Craig Venter was born in Salt Lake City, Utah, on October 14, 1946. He attended the College of San Mateo (a community college) and the University of California at San Diego, from which he received his BS in biochemistry and his PhD in physiology and pharmacology in 1972 and 1975, respectively. He then joined the State University of New York at Buffalo, where he held a variety of posts, also serving concurrently as a cancer researcher at the Roswell Park Memorial Institute. In 1984, he left Buffalo to join the NIH, where he discovered ESTs and their potential for mapping a genome.

In 1992, Venter left the NIH to found his own research institution, The Institute for Genomic Research (TIGR), with the goal of using his own method of gene sequencing (EST) to map the genome of organisms. By 1995, researchers at TIGR had decoded the genome of the first free-living organism, the bacterium *Haemophilus influenzae*. Three years later, Venter founded Celera Genomics for the purpose of sequencing the human genome. He hoped that his approach to mapping would allow him to elucidate the human genome before the federal Human Genome Project could achieve the same goal. Eventually, researchers at Celera and the Human Genome Project began to collaborate in their work and in 2001, they published their results jointly in papers published in the journals *Science* and *Nature*. Since mapping of the human genome has been completed, Celera researchers have also sequenced a number of other genomes, including those of the fruit fly, mouse, rat, cat, dog, African elephant, and common chimpanzee. In June 2005, Venter co-founded Synthetic Genomics, a firm dedicated to the use of genetically modified microorganisms for the production of clean fuels and biochemicals. In March 2016, Venter and his colleagues announced the creation of a synthetic type of bacterium that appears to contain the minimum number of genes needed for life, an organism to which they gave the name JCVI-syn3.0.

James Watson (1928–)

In 1953, Watson and British chemist Francis Crick announced their model for the structure of a DNA molecule. They suggested that the molecule was a double helix—two strands of nucleotides wrapped around each other—consisting of long chains made of alternating phosphate and sugar (deoxyribose) groups, joined to each other by pairs of nitrogen bases. Some historians have called this discovery the most important accomplishment in the history of biology. Since 1953, Watson has devoted the greatest part of his life to administrative assignments, primarily at the Cold Spring Harbor Laboratory, on Long Island, New York. From 1989 to 1992 he was head of the Human Genome Project.

James Dewey Watson was born on April 6, 1928, in Chicago, Illinois. He demonstrated extraordinary intellectual skills at an early age and appeared frequently on a popular radio show of the time called Quiz Kids. He entered the University of Chicago at the age of 15, where he planned to major in his longtime passion of ornithology. After reading Erwin Schrödinger's popular and influential book *What Is Life?*, however, he changed his major to genetics. He received his BS in zoology from Chicago in 1947 and then continued his studies at Indiana University under Salvador Luria, earning his PhD in 1950 at the age of 22.

After completing a postdoctoral year of study at the University of Copenhagen, Watson accepted a second postdoctoral appointment at the world-famous Cavendish Laboratory at the University of Cambridge. There he met Francis Crick and discovered that the two men shared a passion for unraveling the structure of the DNA molecule. The complex series of events that led to their success has been recounted in Watson's book *The Double Helix*.

In 1953, Watson returned to the United States and took a post as senior research fellow at the California Institute of Technology. He then spent another year at Cavendish before accepting an appointment at Harvard University, where he spent the rest of his academic career. He received the 1962 Nobel Prize in Physiology or Medicine, along with Crick, for their discovery of the structure of DNA. He has also received more than two dozen honorary degrees, and a large number of other awards and prizes.

5 Data and Documents

Introduction

One guide to changes that have occurred over the years in DNA technology and its many practical applications is the variety of documents—laws, court cases, reports, and other records—related to the technology. This chapter provides an overview of some of the most important of those documents. Limitations of space prevent the reprinting of complete records, but the excerpts shown here will give a hint of the significance of some of the most important of these documents. The chapter also contains data sets that provide information about quantitative issues related to the development of DNA technology and its applications over the years.

Data

Table 5.1 Adoption of Genetically Engineered Crops in the United States, 1996–2015 (percentage of planted land, in acres)

Year	HT Soybeans	HT Cotton	Bt Cotton	Bt Corn	HT Corn
1996	7	2	15	1	3
1997	17	11	15	8	4
1998	44	26	17	19	9

(continued)

A capillary array used in DNA analysis pictured in a DNA lab of the Oklahoma State Bureau of Investigation in Edmond, Oklahoma. (AP Photo/ Sue Ogrocki)

Table 5.1　(*continued*)

Year	HT Soybeans	HT Cotton	Bt Cotton	Bt Corn	HT Corn
1999	56	42	32	26	8
2000	54	46	35	19	7
2001	68	56	37	19	8
2002	75	58	35	24	11
2003	81	59	41	29	15
2004	85	60	46	33	20
2005	87	61	52	35	26
2006	89	65	57	40	36
2007	91	70	59	49	52
2008	92	68	63	57	63
2009	91	71	65	63	68
2010	93	78	73	63	70
2011	94	73	75	65	72
2012	93	80	77	67	73
2013	93	82	75	76	85
2014	94	91	84	80	89
2015	94	89	84	81	89

HT = herbicide-tolerant.

Bt = plants containing a gene from *Bacillus thuringiensis*.

Source: "Adoption of Genetically Engineered Crops in the United States, 1996–2015." Economic Research Service. U.S. Department of Agriculture. http://www.ers.usda.gov/media/185551/biotechcrops_d.html. Accessed on April 14, 2016.

Table 5.2　Types of Genetically Engineered Corn and Cotton in the United States, 2000–2015 (percentage of planted land, in acres)

Year	Corn			Cotton		
	Bt	Stacked	HT	Bt	Stacked	HT
2000	18	1	6	15	20	26
2001	18	1	7	13	24	32
2002	22	2	9	13	22	36
2003	25	4	11	14	27	32

	Corn			Cotton		
Year	Bt	Stacked	HT	Bt	Stacked	HT
2004	27	6	14	16	30	30
2005	26	8	17	18	34	27
2006	25	15	21	18	29	26
2007	21	28	24	17	42	28
2008	17	40	23	18	45	23
2009	17	46	22	17	48	23
2010	16	47	23	15	58	20
2011	16	49	23	17	58	15
2012	15	52	21	14	63	17
2013	5	71	14	8	67	15
2014	4	76	13	5	79	12
2015	4	77	12	5	79	10

Stacker = BT + Bt.

Source: Adoption of Genetically Engineered Corn/Cotton in the United States, by Trait, 2000–15. Economic Research Service. U.S. Department of Agriculture. http://www.ers.usda.gov/media/186652/biotechcorn_d.html and http://www.ers.usda.gov/media/829782/biotechcotton_d.html. Accessed on April 14, 2016.

Table 5.3   Number of Releases, Sites, and Constructs Authorized by APHIS for Evaluation

Year (FY)	Releases	Sites	Constructs
1989	32	17	74
1990	46	14	142
1991	90	10	226
1992	164	121	427
1993	341	455	870
1994	569	1,669	1,926
1995	734	3,690	2,666
1996	653	2,745	2,305
1997	782	3,427	2,650
1998	1,151	4,781	3,830

(continued)

Table 5.3   (*continued*)

Year (FY)	Releases	Sites	Constructs
1999	1,068	4,134	3,502
2000	1,002	3,836	3,126
2001	1,190	5,831	3,208
2002	1,226	5,111	3,234
2003	824	2,910	2,650
2004	997	4,523	2,851
2005	1011	4,939	3,042
2006	974	4,327	18,532
2007	1,066	3,623	63,217
2008	948	7,744	125,365
2009	846	6,751	217,502
2010	754	6,626	297,422
2011	967	10,128	395,501
2012	767	9,133	469,202

Releases: Permits issued by the Animal and Plant Health Inspections Service (APHIS) of the U.S. Department of Agriculture for field testing of a new product.

Sites: Locations of field testing.

Construct: Unit required for the expression of a specific gene in the field.

Source: Fernandez-Cornejo, Jorge, et al. 2014. "Genetically Engineered Crops in the United States." ERR-162. U.S. Department of Agriculture. Economic Research Service, Table 1, page 5. Available online at http://www.ers.usda.gov/media/1282246/err162.pdf. Accessed on April 15, 2016.

Table 5.4   Number of Releases of Genetically Engineered Varieties by APHIS, by Crop, as of September 2013

Crop	Approved Releases
Corn	7,778
Soybeans	2,225
Cotton	1,104
Potato	904
Tomato	688
Wheat	485

Crop	Approved Releases
Alfalfa	452
Tobacco	427
Rapeseed	310
Rice	294

Source: Fernandez-Cornejo, Jorge, et al. 2014. "Genetically Engineered Crops in the United States." ERR-162. U.S. Department of Agriculture. Economic Research Service, Figure 2, page 6. Available online at http://www.ers.usda.gov/media/1282246/err162.pdf. Accessed on April 15, 2016.

Table 5.5   Number of Releases Approved by APHIS by Gene Trait, to September 2013

Trait	Number of Releases
Herbicide tolerance	6,772
Agronomic properties	5,190
Product quality	4,896
Insect resistance	4,809
Marker gene	1,892
Virus resistance	1,425
Fungal resistance	1,191
Bacterial resistance	224
Nematode resistance	149
Other	1,986

Source: Fernandez-Cornejo, Jorge, et al. 2014. "Genetically Engineered Crops in the United States." ERR-162. U.S. Department of Agriculture. Economic Research Service, Figure 3, page 6. Available online at http://www.ers.usda.gov/media/1282246/err162.pdf. Accessed on April 15, 2016.

Table 5.6   Institutions with Greatest Number of APHIS Permits for Genetically Engineered Crops, to September 2013

Institution	Number of Permits
Monsanto	6,782
Pioneer (now part of DuPont)	1,085
Syngenta	565

(continued)

Table 5.6   (*continued*)

Institution	Number of Permits
Agriculture Research Service, USDA	370
AgrEvo	326
Dow AgroSciences	400
DuPont	320
ArborGen	311
Bayer CropScience	260
Seminis	210

Source: Fernandez-Cornejo, Jorge, et al. 2014. "Genetically Engineered Crops in the United States." ERR-162. U.S. Department of Agriculture. Economic Research Service, Figure 4, page 7. Available online at http://www.ers.usda.gov/media/1282246/err162.pdf. Accessed on April 15, 2016.

Table 5.7   CODIS Statistics for 15 Top States, as of February 2016

State	Offender Profiles	Forensic Profiles	Investigations Aided
California	1,817,352	70,528	47,507
Florida	988,409	57,900	31,550
Texas	791,385	60,175	23,818
New York	554,287	51,420	20,409
Illinois	550,264	36,078	19,926
Ohio	462,165	49,068	15,050
Missouri	281,576	20,972	11,059
Michigan	366,157	23,107	10,411
Virginia	396,262	19,094	9,541
New Jersey	289,030	18,690	9,142
Arizona	301,273	18,078	8,235
Alabama	232,188	14,250	6,717
Colorado	178,871	14,291	6,655
South Carolina	191,558	13,242	6,402
Pennsylvania	339,831	15,243	6,399

Source: CODIS—NDIS Statistics. 2016. Federal Bureau of Investigation. https://www.fbi.gov/about-us/lab/biometric-analysis/codis/ndis-statistics. Accessed on April 15, 2016.

Documents

Daubert v. Merrell Dow Pharmaceuticals, Inc. (1993)

A crucial issue in many court cases is how a judge and/or jury is to interpret "scientific" evidence that is submitted for consideration. How do such individuals, usually with no special training in science, know that such evidence is true and reliable? For many years in the United States, the basis for making that decision was the so-called Frye standard, elucidated in a 1923 decision by the Court of Appeals for the District of Columbia. That court said that scientific evidence was admissible if it was "generally accepted" by the scientific community, a standard that gradually evolved in specificity over time. In 1993, the U.S. Supreme Court said that the Frye standard was no longer sufficiently precise, and it established a new standard for admissibility, now called the Daubert standard, after the case in which it was announced, Daubert v. Merrell Dow Pharmaceuticals, Inc. *The essence of the Court's ruling was as follows (omitted citations are indicated with triple asterisks, ***):*

The Frye test has its origin in a short and citation-free 1923 decision concerning the admissibility of evidence derived from a systolic blood pressure deception test, a crude precursor to the polygraph machine. In what has become a famous (perhaps infamous) passage, the then Court of Appeals for the District of Columbia described the device and its operation and declared:

> "Just when a scientific principle or discovery crosses the line between the experimental and demonstrable stages *** is difficult to define. Somewhere in this twilight zone the evidential force of the principle must be recognized, and while courts will go a long way in admitting expert testimony deduced from a well-recognized scientific principle or discovery, the thing from which the deduction is made must be sufficiently established to have gained general acceptance in the particular field in which it belongs."

Because the deception test had "not yet gained such standing and scientific recognition among physiological and psychological authorities as would justify the courts in admitting expert testimony deduced from the discovery, development, and experiments thus far made," evidence of its results was ruled inadmissible. ***

. . .

The major portion of the Court's decision, then, is based on the observation that an extensive and detailed standard for evidence now exists in the United States, the Federal Rules of Evidence, one section of which (702) deals with just the issue the Court is dealing with here. It determines that Rule 702 is a better standard for the admissibility of scientific evidence and explains in detail why that it is, before concluding that:

To summarize: "General acceptance" is not a necessary precondition to the admissibility of scientific evidence under the Federal Rules of Evidence, but the Rules of Evidence—especially Rule 702—do assign to the trial judge the task of ensuring that an expert's testimony both rests on a reliable foundation and is relevant to the task at hand. Pertinent evidence based on scientifically valid principles will satisfy those demands.

The inquiries of the District Court and the Court of Appeals focused almost exclusively on "general acceptance," as gauged by publication and the decisions of other courts. Accordingly, *** the judgment of the Court of Appeals is vacated, and the case is remanded for further proceedings consistent with this opinion.

Source: Daubert v. Merrell Dow Pharmaceuticals, Inc., 509 U.S. 579. 1993.

Executive Order 13145 (2000)

Federal policy on the use of information obtained from genetic testing is enshrined in two documents, a presidential executive order signed by President Bill Clinton on February 8, 2000, and the Genetic Information Nondiscrimination Act of 2008 (see the description that follows). Clinton's order lays out federal policy on the use of genetic information, describes the type of information covered in the order, and lists certain exceptions to the requirements posed in the order. The most relevant parts of the order are as follows:

Section 1. Nondiscrimination in Federal Employment on the Basis of Protected Genetic Information.

> 1-101. It is the policy of the Government of the United States to provide equal employment opportunity in Federal employment for all qualified persons and to prohibit discrimination against employees based on protected genetic information, or information about a request for or the receipt of genetic services. This policy of equal opportunity applies to every aspect of Federal employment.

> 1-102. The head of each Executive department and agency shall extend the policy set forth in section 1-101 to all its employees covered by section 717 of Title VII of the Civil Rights Act of 1964, as amended (42 U.S.C. 2000e-16).

> 1-103. Executive departments and agencies shall carry out the provisions of this order to the extent permitted by law and consistent with their statutory and regulatory authorities, and their enforcement mechanisms. The Equal Employment Opportunity Commission shall be responsible for coordinating the policy of the Government of the United States to prohibit discrimination against employees in Federal employment based on protected genetic information, or information about a request for or the receipt of genetic services.

* . . . [Section 2 lists definitions of terms used in the order and provides specific nondiscrimination responsibilities for departments and agencies. Section 3 lists exceptions allowed to the execution of this order. Section 4 deals with miscellaneous issues.]

Source: "Executive Order: February 8, 2000." 65 Federal Register 6877, February 10, 2000.

Regulations with Respect to Genetically Modified Foods: European Union (2003)

In 1997, the European Parliament adopted a set of rules and regulations controlling the use of genetically modified foods in the European Community (EC No. 258/97). In 2003, the parliament adopted a number of revisions to those rules, covering the labeling and sale of foods and feed in Europe (EC No. 1829/2003). The directive is divided into four chapters. Chapter I deals with definitions and objectives of the act. Chapter II deals with genetically modified foods. Chapter III deals with genetically modified feeds. Chapter IV deals with additional administrative and regulatory provisions relating to either GM foods or feed, or both. Some sections of those rules are excerpted here.

Chapter II

Section 1

ARTICLE 4

Requirements

1. Food referred to in Article 3(1) must not:
   (a) have adverse effects on human health, animal health or the environment;
   (b) mislead the consumer;
   (c) differ from the food which it is intended to replace to such an extent that its normal consumption would be nutritionally disadvantageous for the consumer.

2. No person shall place on the market a GMO for food use or food referred to in Article 3(1) unless it is covered by an authorisation granted in accordance with this Section and the relevant conditions of the authorisation are satisfied.

3. No GMO for food use or food referred to in Article 3(1) shall be authorised unless the applicant for such authorisation has adequately and sufficiently demonstrated that it satisfies the requirements of paragraph 1 of this Article.

*... *[The rest of this article describes how an authorization is obtained for a GM food. Remaining articles in Section 1 (Articles 5–11) discuss further details about authorizing the sale of GM foods.]*

Section 2

LABELLING

Article 12

Scope

1. This Section shall apply to foods which are to be delivered as such to the final consumer or mass caterers in the Community and which:

   (a) contain or consist of GMOs; or

   (b) are produced from or contain ingredients produced from GMOs.

2. This Section shall not apply to foods containing material which contains, consists of or is produced from GMOs in a proportion no higher than 0,9 per cent of the food ingredients considered individually or food consisting of a single ingredient, provided that this presence is adventitious or technically unavoidable.

*...

Article 13

Requirements

1. Without prejudice to the other requirements of Community law concerning the labelling of foodstuffs, foods

falling within the scope of this Section shall be subject to the following specific labelling requirements:

(a) where the food consists of more than one ingredient, the words 'genetically modified' or 'produced from genetically modified (name of the ingredient)' shall appear in the list of ingredients provided for in Article 6 of Directive 2000/13/EC in parentheses immediately following the ingredient concerned;

(b) where the ingredient is designated by the name of a category, the words 'contains genetically modified (name of organism)' or 'contains (name of ingredient) produced from genetically modified (name of organism)' shall appear in the list of ingredients;

(c) where there is no list of ingredients, the words 'genetically modified' or 'produced from genetically modified (name of organism)' shall appear clearly on the labelling;

(d) the indications referred to in (a) and (b) may appear in a footnote to the list of ingredients. In this case they shall be printed in a font of at least the same size as the list of ingredients. Where there is no list of ingredients, they shall appear clearly on the labelling;

(e) where the food is offered for sale to the final consumer as non-pre-packaged food, or as pre-packaged food in small containers of which the largest surface has an area of less than 10 cm^2, the information required under this paragraph must be permanently and visibly displayed either on the food display or immediately next to it, or on the packaging material, in a font sufficiently large for it to be easily identified and read.* . . .

[Article 14 deals with more details about the labeling of products containing GM foods.]

Source: "Regulation (EC) No 1829/2003 of the European Parliament and of the Council of 22 September 2003 on Genetically Modified Food and Feed." *Official Journal of the European Union.* 268 (October 18, 2003): 1–23. ©European Communities, http://ec.europa.eu/food/food/animalnutrition/labelling/Reg_1829_2003_en.pdf. Only European Union legislation printed in the paper edition of the *Official Journal of the European Union* is deemed authentic. Accessed on April 14, 2016.

Alaska State Law on Genetic Privacy (2004)

Until 2008, there were only limited federal regulations dealing with genetic testing. By the time the first law was passed by the U.S. Congress (see the next document), a number of states had adopted legislation regulating genetic testing in one way or another. The Alaska law, excerpted here, is an example of one way states have dealt with the issue of genetic testing.

Chapter 18.13. GENETIC PRIVACY
   Sec. 18.13.010. Genetic testing.

(a) Except as provided in (b) of this section,

   (1) a person may not collect a DNA sample from a person, perform a DNA analysis on a sample, retain a DNA sample or the results of a DNA analysis, or disclose the results of a DNA analysis unless the person has first obtained the informed and written consent of the person, or the person's legal guardian or authorized representative, for the collection, analysis, retention, or disclosure;

   (2) a DNA sample and the results of a DNA analysis performed on the sample are the exclusive property of the person sampled or analyzed.

(b) The prohibitions of (a) of this section do not apply to DNA samples collected and analyses conducted

(1) under AS 44.41.035 or comparable provisions of another jurisdiction;

(2) for a law enforcement purpose, including the identification of perpetrators and the investigation of crimes and the identification of missing or unidentified persons or deceased individuals;

(3) for determining paternity;

(4) to screen newborns as required by state or federal law;

(5) for the purpose of emergency medical treatment.

(c) A general authorization for the release of medical records or medical information may not be construed as the informed and written consent required by this section. The Department of Health and Social Services may by regulation adopt a uniform informed and written consent form to assist persons in meeting the requirements of this section. A person using that uniform informed and written consent is exempt from civil or criminal liability for actions taken under the consent form. A person may revoke or amend their informed and written consent at any time.

Sec. 18.13.020. Private right of action.

A person may bring a civil action against a person who collects a DNA sample from the person, performs a DNA analysis on a sample, retains a DNA sample or the results of a DNA analysis, or discloses the results of a DNA analysis in violation of this chapter. In addition to the actual damages suffered by the person, a person violating this chapter shall be liable to the person for damages in the amount of $5,000 or, if the violation resulted in profit or monetary gain to the violator, $100,000.

Sec. 18.13.030. Criminal penalty.

(a) A person commits the crime of unlawful DNA collection, analysis, retention, or disclosure if the person knowingly collects a DNA sample from a person, performs a DNA analysis on a sample, retains a DNA sample or the results of a DNA analysis, or discloses the results of a DNA analysis in violation of this chapter.

(b) In this section, "knowingly" has the meaning given in AS 11.81.900.

(c) Unlawful DNA collection, analysis, retention, or disclosure is a class A misdemeanor.

Source: Alaska Statutes 2015. http://www.legis.state.ak.us/basis/statutes.asp#18.13.010. Accessed on April 15, 2016.

Post-Conviction DNA Testing (2004)

One of the most important contributions that DNA testing has made to forensic science and law enforcement is its use in determining the possible innocence of individuals who were convicted of a crime at sometime in the past, but have since insisted upon their innocence. DNA testing that may not have been available or that was not used following the original crime may provide new evidence in considering such an individual's plea for reconsideration of the original verdict. A number of organizations, most prominently the Innocence Project, have long lobbied for state and federal legislation that would make it possible and/or easier for a person to request DNA testing that might provide his or her exoneration for a crime. In 2004, President George W. Bush signed the Justice for All Act that included a long section providing conditions for such a situation. Some major elements of that legislation are as follows:

(a) In General.—Upon a written motion by an individual under a sentence of imprisonment or death pursuant to a conviction for a Federal offense (referred to in this section as the "applicant"), the court that entered the judgment of conviction shall order DNA testing of specific evidence if the court finds that all of the following apply:

(1) The applicant asserts, under penalty of perjury, that the applicant is actually innocent of—

(A) the Federal offense for which the applicant is under a sentence of imprisonment or death; or

(B) another Federal or State offense, if—

    (i) evidence of such offense was admitted during a Federal death sentencing hearing and exoneration of such offense would entitle the applicant to a reduced sentence or new sentencing hearing; and

    (ii) in the case of a State offense—

        (I) the applicant demonstrates that there is no adequate remedy under State law to permit DNA testing of the specified evidence relating to the State offense; and

        (II) to the extent available, the applicant has exhausted all remedies available under State law for requesting DNA testing of specified evidence relating to the State offense.

(2) The specific evidence to be tested was secured in relation to the investigation or prosecution of the Federal or State offense referenced in the applicant's assertion under paragraph (1).

(3) The specific evidence to be tested—

(A) was not previously subjected to DNA testing and the applicant did not—

    (i) knowingly and voluntarily waive the right to request DNA testing of that evidence in a court proceeding after the date of enactment of the Innocence Protection Act of 2004; or

    (ii) knowingly fail to request DNA testing of that evidence in a prior motion for postconviction DNA testing; or

(B) was previously subjected to DNA testing and the applicant is requesting DNA testing using a new method or technology that is substantially more probative than the prior DNA testing.

(4) The specific evidence to be tested is in the possession of the Government and has been subject to a chain of custody and retained under conditions sufficient to ensure that such evidence has not been substituted, contaminated, tampered with, replaced, or altered in any respect material to the proposed DNA testing.

(5) The proposed DNA testing is reasonable in scope, uses scientifically sound methods, and is consistent with accepted forensic practices.

(6) The applicant identifies a theory of defense that—

(A) is not inconsistent with an affirmative defense presented at trial; and

(B) would establish the actual innocence of the applicant of the Federal or State offense referenced in the applicant's assertion under paragraph (1).

(7) If the applicant was convicted following a trial, the identity of the perpetrator was at issue in the trial.

(8) The proposed DNA testing of the specific evidence may produce new material evidence that would—

(A) support the theory of defense referenced in paragraph (6); and

(B) raise a reasonable probability that the applicant did not commit the offense.

(9) The applicant certifies that the applicant will provide a DNA sample for purposes of comparison.

(10) The motion is made in a timely fashion, subject to the following conditions:

(A) There shall be a rebuttable presumption of timeliness if the motion is made within 60 months of enactment of the Justice For All Act of 2004 or within 36 months of conviction, whichever comes later. Such presumption may be rebutted upon a showing—

(i) that the applicant's motion for a DNA test is based solely upon information used in a previously denied motion; or

(ii) of clear and convincing evidence that the applicant's filing is done solely to cause delay or harass.

(B) There shall be a rebuttable presumption against timeliness for any motion not satisfying subparagraph (A) above. Such presumption may be rebutted upon the court's finding—

(i) that the applicant was or is incompetent and such incompetence substantially contributed to the delay in the applicant's motion for a DNA test;

(ii) the evidence to be tested is newly discovered DNA evidence;

(iii) that the applicant's motion is not based solely upon the applicant's own assertion of innocence and, after considering all relevant facts and circumstances surrounding the motion, a denial would result in a manifest injustice; or

(iv) upon good cause shown.

(C) *[A list of definitions is provided here.]*

(b) Notice to the Government; Preservation Order; Appointment of Counsel.—

(c) Testing Procedures.—

(d) Time Limitation in Capital Cases.

(e) Reporting of Test Results.—

[These sections describe the procedural elements required for DNA testing process.]

(f) Post-Testing Procedures; Inconclusive and Inculpatory Results.—

 (1) Inconclusive results.—

  If DNA test results obtained under this section are inconclusive, the court may order further testing, if appropriate, or may deny the applicant relief.

 (2) Inculpatory results.—If DNA test results obtained under this section show that the applicant was the source of the DNA evidence, the court shall—

 (A) deny the applicant relief; and

 (B) on motion of the Government—

  (i) make a determination whether the applicant's assertion of actual innocence was false, and, if the court makes such a finding, the court may hold the applicant in contempt;

  (ii) assess against the applicant the cost of any DNA testing carried out under this section;

  (iii) forward the finding to the Director of the Bureau of Prisons, who, upon receipt of such a finding, may deny, wholly or in part, the good conduct credit authorized under section 3632 on the basis of that finding;

  (iv) if the applicant is subject to the jurisdiction of the United States Parole Commission, forward the finding to the Commission so that the Commission may deny parole on the basis of that finding; and

  (v) if the DNA test results relate to a State offense, forward the finding to any appropriate State official.

(3) Sentence.—

In any prosecution of an applicant under chapter 79 for false assertions or other conduct in proceedings under this section, the court, upon conviction of the applicant, shall sentence the applicant to a term of imprisonment of not less than 3 years, which shall run consecutively to any other term of imprisonment the applicant is serving.

(g) Post-Testing Procedures; Motion for New Trial or Resentencing.—

(1) In general.—

Notwithstanding any law that would bar a motion under this paragraph as untimely, if DNA test results obtained under this section exclude the applicant as the source of the DNA evidence, the applicant may file a motion for a new trial or resentencing, as appropriate. The court shall establish a reasonable schedule for the applicant to file such a motion and for the Government to respond to the motion.

(2) Standard for granting motion for new trial or resentencing.—The court shall grant the motion of the applicant for a new trial or resentencing, as appropriate, if the DNA test results, when considered with all other evidence in the case (regardless of whether such evidence was introduced at trial), establish by compelling evidence that a new trial would result in an acquittal of—

(A) in the case of a motion for a new trial, the Federal offense for which the applicant is under a sentence of imprisonment or death; and

(B) in the case of a motion for resentencing, another Federal or State offense, if evidence of such offense was admitted during a Federal death sentencing hearing and exoneration of such offense would entitle the applicant to a reduced sentence or a new sentencing proceeding.

(h) Other Laws Unaffected.—

[This section discusses other laws that may be related to the DNA testing request.]

Source: Chapter 228A—Post-Conviction DNA Testing. 2016. Office of the Law Revision Council. United States Code, Title 18, pages 234–238.

United Nations Declaration on Human Cloning (2005)

The announcement of the first cloned mammal (the sheep, Dolly) in 1997 fueled a firestorm of concern about the possible use of DNA technology to clone a human being. One response evoked by this event was the request by some members of the United Nations General Assembly for a study of this issue and a statement as to the UN's position on human reproductive cloning. The first formal request for such an action was presented in 2001, and a committee to study the issue and publish its report was appointed by the General Assembly in that year. Committee negotiations did not go smoothly for a number of reasons, most important of which was that some nations wished to ban all forms of cloning, including both therapeutic and reproductive cloning, while other nations preferred a limited ban on reproductive cloning only. This division was reflected in the final vote on the declaration adopted by the General Assembly on March 8, 2005, by a vote of 84 to 34, with 37 abstentions and 36 members being absent from the vote. The text of that declaration follows.

United Nations Declaration on Human Cloning

The General Assembly,

Guided by the purposes and principles of the Charter of the United Nations,

Recalling the Universal Declaration on the Human Genome and Human Rights, adopted by the General Conference of the United Nations Educational, Scientific and Cultural

Organization on 11 November 1997, and in particular article 11 thereof, which states that practices which are contrary to human dignity, such as the reproductive cloning of human beings, shall not be permitted,

Recalling also its resolution 53/152 of 9 December 1998, by which it endorsed the Universal Declaration on the Human Genome and Human Rights,

Aware of the ethical concerns that certain applications of rapidly developing life sciences may raise with regard to human dignity, human rights and the fundamental freedoms of individuals,

Reaffirming that the application of life sciences should seek to offer relief from suffering and improve the health of individuals and humankind as a whole,

Emphasizing that the promotion of scientific and technical progress in life sciences should be sought in a manner that safeguards respect for human rights and the benefit of all,

Mindful of the serious medical, physical, psychological and social dangers that human cloning may imply for the individuals involved, and also conscious of the need to prevent the exploitation of women,

Convinced of the urgency of preventing the potential dangers of human cloning to human dignity,

Solemnly declares the following:

(a) Member States are called upon to adopt all measures necessary to protect adequately human life in the application of life sciences;

(b) Member States are called upon to prohibit all forms of human cloning inasmuch as they are incompatible with human dignity and the protection of human life;

(c) Member States are further called upon to adopt the measures necessary to prohibit the application of genetic engineering techniques that may be contrary to human dignity;

(d) Member States are called upon to take measures to prevent the exploitation of women in the application of life sciences;

(e) Member States are also called upon to adopt and implement without delay national legislation to bring into effect paragraphs (a) to (d);

(f) Member States are further called upon, in their financing of medical research, including of life sciences, to take into account the pressing global issues such as HIV/AIDS, tuberculosis and malaria, which affect in particular the developing countries.

Source: United Nations General Assembly. "International Convention against the Reproductive Cloning of Human Beings." A/59/516/Add. 1, February 24, 2005. ©United Nations, 2005. Reproduced with permission.

Additional Protocol to the Convention on Human Rights and Biomedicine, Concerning Genetic Testing for Health Purposes (2008)

This document has a long life, dating back to the late 1990s when the Steering Committee on Bioethics of the Council of Europe began deliberations on the use of genetic testing in the diagnosis of human disease. Those deliberations became very complex, and the committee's final report was not issued until almost a decade later. The final protocol consists of 11 chapters and 28 articles, whose major foci are as follows:

Chapter I—Object and scope

Article 1—Object and Purpose
Parties to this Protocol shall protect the dignity and identity of all human beings and guarantee everyone, without discrimination, respect for their integrity and other rights and fundamental freedoms with regard to the tests to which this Protocol applies in accordance with Article 2.

Article 2—Scope

1   This Protocol applies to tests, which are carried out for health purposes, involving analysis of biological samples of human origin and aiming specifically to identify the genetic characteristics of a person which are inherited or acquired during early prenatal development (hereinafter referred to as "genetic tests").

2   This Protocol does not apply:

   a to genetic tests carried out on the human embryo or foetus;

   b to genetic tests carried out for research purposes. * . . .

Chapter II—General provisions

Article 3—Primacy of the Human Being

The interests and welfare of the human being concerned by genetic tests covered by this Protocol shall prevail over the sole interest of society or science.

Article 4—Non-Discrimination and Non-Stigmatisation

1   Any form of discrimination against a person, either as an individual or as a member of a group on grounds of his or her genetic heritage is prohibited.

2   Appropriate measures shall be taken in order to prevent stigmatisation of persons or groups in relation to genetic characteristics.

Chapter III—Genetic Services

Article 5—Quality of Genetic Services

Parties shall take the necessary measures to ensure that genetic services are of appropriate quality. In particular, they shall see to it that:

a   genetic tests meet generally accepted criteria of scientific validity and clinical validity;

b   a quality assurance programme is implemented in each laboratory and that laboratories are subject to regular monitoring;

c   persons providing genetic services have appropriate qualifications to enable them to perform their role in accordance with professional obligations and standards.

Article 6—Clinical utility

Clinical utility of a genetic test shall be an essential criterion for deciding to offer this test to a person or a group of persons.

Article 7—Individualised supervision

1   A genetic test for health purposes may only be performed under individualised medical supervision.

2   Exceptions to the general rule referred to in paragraph 1 may be allowed by a Party, subject to appropriate measures being provided, taking into account the way the test will be carried out, to give effect to the other provisions of this Protocol. However, such an exception may not be made with regard to genetic tests with important implications for the health of the persons concerned or members of their family or with important implications concerning procreation choices.

Chapter IV—Information, Genetic Counselling and Consent

Article 8—Information and Genetic Counselling

1   When a genetic test is envisaged, the person concerned shall be provided with prior appropriate information in particular on the purpose and the nature of the test, as well as the implications of its results.

2   For predictive genetic tests as referred to in Article 12 of the Convention on Human Rights and Biomedicine, appropriate genetic counselling shall also be available for the person concerned.

The tests concerned are:

- tests predictive of a monogenic disease,
- tests serving to detect a genetic predisposition or genetic susceptibility to a disease,
- tests serving to identify the subject as a healthy carrier of a gene responsible for a disease.

The form and extent of this genetic counselling shall be defined according to the implications of the results of the test and their significance for the person or the members of his or her family, including possible implications concerning procreation choices.

Genetic counselling shall be given in a non-directive manner.

Article 9—Consent

1   A genetic test may only be carried out after the person concerned has given free and informed consent to it. Consent to tests referred to in Article 8, paragraph 2, shall be documented.

2   The person concerned may freely withdraw consent at any time.

*. . . [Chapter V deals with genetic testing of individuals unable to give their consent. Chapter VI deals with genetic testing of family members who are unable to give consent, who cannot be contacted, or who are deceased.]

Chapter VII—Private Life and Right to Information

Article 16—Respect for Private Life and Right to Information

1 Everyone has the right to respect for his or her private life, in particular to protection of his or her personal data derived from a genetic test.

2 Everyone undergoing a genetic test is entitled to know any information collected about his or her health derived from this test. The conclusions drawn from the test shall be accessible to the person concerned in a comprehensible form.

3 The wish of a person not to be informed shall be respected.

4 In exceptional cases, restrictions may be placed by law on the exercise of the rights contained in paragraphs 2 and 3 above in the interests of the person concerned.

Chapter VIII—Genetic Screening Programmes for Health Purposes

Article 19—Genetic Screening Programmes for Health Purposes

A health screening programme involving the use of genetic tests may only be implemented if it has been approved by the competent body. This approval may only be given after independent evaluation of its ethical acceptability and fulfilment of the following specific conditions:

a the programme is recognised for its health relevance for the whole population or section of population concerned;

b the scientific validity and effectiveness of the programme have been established;

c appropriate preventive or treatment measures in respect of the disease or disorder which is the subject of the screening, are available to the persons concerned;

d   appropriate measures are provided to ensure equitable access to the programme; e the programme provides measures
to adequately inform the population or section of population concerned of the existence, purposes and means of ac-
cessing the screening programme as well as the voluntary
nature of participation in it.

*. . . [Chapter IX deals with public information programs about
genetic testing. Chapter X deals with the relation between this protocol and the general Convention. Chapter XI deals with "housekeeping" issues needed to carry out terms of the protocol.]

Source: "Additional Protocol to the Convention on Human
Rights and Biomedicine, Concerning Genetic Testing for Health
Purposes. Strasbourg: Council of Europe, November 27, 2008.
Available online. http://conventions.coe.int/Treaty/EN/Treaties/
Html/203.htm. Accessed on April 16, 2016.

Genetic Information Nondiscrimination Act (2008)

On January 16, 2007, Representative Louise Slaughter (D-NY)
introduced in the U.S. House of Representatives legislation to
protect individuals from discrimination based on personal information obtained as the result of genetic testing. That legislation
worked its way through the U.S. Congress and was signed by President George W. Bush on May 21, 2008, becoming Public Law
110–233. The new law amends a number of existing laws dealing with employment, taxation, retirement, and related issues, and
also establishes a number of new regulations about the use of genetic testing by employers. Some important sections of that law
follow.

Sec. 101. Amendments to Employee Retirement Income Security Act of 1974

(a) NO DISCRIMINATION IN GROUP PREMIUMS
BASED ON GENETIC INFORMATION.—

*
. . .

"(A) IN GENERAL.—For purposes of this section, a group health plan, and a health insurance issuer offering group health insurance coverage in connection with a group health plan, may not adjust premium or contribution amounts for the group covered under such plan on the basis of genetic information.

*
. . .

(b) LIMITATIONS ON GENETIC TESTING; PROHIBITION ON COLLECTION OF GENETIC INFORMATION; APPLICATION TO ALL PLANS.

*
. . .

TESTING.—A group health plan, and a health insurance issuer offering health insurance coverage in connection with a group health plan, shall not request or require an individual or a family member of such individual to undergo a genetic test.

*
. . .

"(d) PROHIBITION ON COLLECTION OF GENETIC INFORMATION.—

"(1) IN GENERAL.—A group health plan, and a health insurance issuer offering health insurance coverage in connection with a group health plan, shall not request, require, or purchase genetic information for underwriting purposes (as defined in section 733).

*. . . *[Section 102 deals with amendments to the Public Health Service Act. Section 103 deals with amendments to the Internal Revenue Service code. Section 104 deals with amendments to the Social Service Act. Sections 105 and 106 are "housekeeping" sections dealing with privacy, confidentiality, and related matters.]*

Title II—Prohibiting Employment Discrimination On the Basis of Genetic Information
[Section 201 supplies definitions of terms used in this title.]

SEC. 202. Employer Practices

(a) DISCRIMINATION BASED ON GENETIC INFOR-MATION.—It shall be an unlawful employment practice for an employer—

  (1) to fail or refuse to hire, or to discharge, any employee, or otherwise to discriminate against any employee with respect to the compensation, terms, conditions, or privileges of employment of the employee, because of genetic information with respect to the employee; or

  (2) to limit, segregate, or classify the employees of the employer in any way that would deprive or tend to deprive any employee of employment opportunities or otherwise adversely affect the status of the employee as an employee, because of genetic information with respect to the employee.

(b) ACQUISITION OF GENETIC INFORMATION.—It shall be an unlawful employment practice for an employer to request, require, or purchase genetic information with respect to an employee or a family member of the employee except—

  (1) where an employer inadvertently requests or requires family medical history of the employee or family member of the employee;

  (2) where—

    (A) health or genetic services are offered by the employer, including such services offered as part of a wellness program;

    (B) the employee provides prior, knowing, voluntary, and written authorization;

    (C) only the employee (or family member if the family member is receiving genetic services) and the licensed health care professional or board certified genetic counselor involved in providing such services receive individually identifiable information concerning the results of such services; and

(D) any individually identifiable genetic information provided under subparagraph (C) in connection with the services provided under subparagraph (A) is only available for purposes of such services and shall not be disclosed to the employer except in aggregate terms that do not disclose the identity of specific employees;

(3) where an employer requests or requires family medical history from the employee to comply with the certification provisions of section 103 of the Family and Medical Leave Act of 1993 (29 U.S.C. 2613) or such requirements under State family and medical leave laws;

(4) where an employer purchases documents that are commercially and publicly available (including newspapers, magazines, periodicals, and books, but not including medical databases or court records) that include family medical history;

(5) where the information involved is to be used for genetic monitoring of the biological effects of toxic substances in the workplace, but only if—

(A) the employer provides written notice of the genetic monitoring to the employee;

(B)

(i) the employee provides prior, knowing, voluntary, and written authorization; or

(ii) the genetic monitoring is required by Federal or State law;

(C) the employee is informed of individual monitoring results;

(D) the monitoring is in compliance with—

   (i) any Federal genetic monitoring regulations, including any such regulations that may be promulgated by the Secretary of Labor pursuant to the Occupational Safety and Health Act of 1970 (29 U.S.C.651 et seq.), the Federal Mine Safety and Health Act of 1977 (30 U.S.C. 801 et seq.), or the Atomic Energy Act of 1954 (42 U.S.C. 2011 et seq.); or

   (ii) State genetic monitoring regulations, in the case of a State that is implementing genetic monitoring regulations under the authority of the Occupational Safety and Health Act of 1970 (29 U.S.C. 651 et seq.); and

(E) the employer, excluding any licensed health care professional or board certified genetic counselor that is involved in the genetic monitoring program, receives the results of the monitoring only in aggregate terms that do not disclose the identity of specific employees; or

(6) where the employer conducts DNA analysis for law enforcement purposes as a forensic laboratory or for purposes of human remains identification, and requests or requires genetic information of such employer's employees, but only to the extent that such genetic information is used for analysis of DNA identification markers for quality control to detect sample contamination.

. . . [Section 203 deals with requirements of employment agencies with regard to genetic testing.]

Sec. 204. Labor Organization Practices

(a) DISCRIMINATION BASED ON GENETIC INFORMATION.—It shall be an unlawful employment practice for a labor organization—

(1) to exclude or to expel from the membership of the organization, or otherwise to discriminate against, any
member because of genetic information with respect
to the member;

(2) to limit, segregate, or classify the members of the organization, or fail or refuse to refer for employment
any member, in any way that would deprive or tend
to deprive any member of employment opportunities,
or otherwise adversely affect the status of the member
as an employee, because of genetic information with
respect to the member; or

(3) to cause or attempt to cause an employer to discriminate against a member in violation of this title.

*. . . [The rest of this section provides restrictions on unions simi-
lar to those applied to employers in section 201. Section 205 deals
with nondiscrimination in training programs. Section 206 deals
with confidentiality of genetic information. Sections 207 through
210 and Title III deal with miscellaneous provisions and "housekeeping" issues related to enactment and enforcement of the law.]*

Source: Public Law 110–233—May 21, 2008.

Vermont Act 120 (GM Food Labeling) (2014)

*As of mid-2016, only one state has adopted a law requiring the
labeling of foods that contain genetically modified materials. That
law went into effect on July 1, 2016. The following excerpts are
taken from that act.*

§ 3043. Labeling of Food Produced with
Genetic Engineering

(a) Except as set forth in section 3044 of this title, food offered for sale by a retailer after July 1, 2016 shall be labeled
as produced entirely or in part from genetic engineering if
it is a product:

(1) offered for retail sale in Vermont; and

(2) entirely or partially produced with genetic engineering.

(b) If a food is required to be labeled under subsection (a) of this section, it shall be labeled as follows:

(1) in the case of a packaged raw agricultural commodity, the manufacturer shall label the package offered for retail sale, with the clear and conspicuous words "produced with genetic engineering";

(2) in the case of any raw agricultural commodity that is not separately packaged, the retailer shall post a label appearing on the retail store shelf or bin in which the commodity is displayed for sale with the clear and conspicuous words "produced with genetic engineering"; or

(3) in the case of any processed food that contains a product or products of genetic engineering, the manufacturer shall label the package in which the processed food is offered for sale with the words: "partially produced with genetic engineering"; "may be produced with genetic engineering"; or "produced with genetic engineering."

(c) Except as set forth under section 3044 of this title, a manufacturer of a food produced entirely or in part from genetic engineering shall not label the product on the package, in signage, or in advertising as "natural," "naturally made," "naturally grown," "all natural," or any words of similar import that would have a tendency to mislead a consumer.

(d) This section and the requirements of this chapter shall not be construed to require:

(1) the listing or identification of any ingredient or ingredients that were genetically engineered; or

(2) the placement of the term "genetically engineered" immediately preceding any common name or primary product descriptor of a food.

Section 3044 deals with exceptions.

§ 3045. Retailer Liability

(a) A retailer shall not be liable for the failure to label a processed food as required by section 3043 of this title, unless the retailer is the producer or manufacturer of the processed food.

(b) A retailer shall not be held liable for failure to label a raw agricultural commodity as required by section 3043 of this title, provided that the retailer, within 30 days of any proposed enforcement action or notice of violation, obtains a sworn statement in accordance with subdivision 3044(2) of this title.

[Section 3046 deals with severability.
Section 3047 deals with false certification.]

§ 3048. Penalties; Enforcement

(a) Any person who violates the requirements of this chapter shall be liable for a civil penalty of not more than $1,000.00 per day, per product. Calculation of the civil penalty shall not be made or multiplied by the number of individual packages of the same product displayed or offered for retail sale. Civil penalties assessed under this section shall accrue and be assessed per each uniquely named, designated, or marketed product.

Source: "No. 120. An Act Relating to the Labeling of Food Produced with Genetic Engineering." Vermont General Assembly. http://www.leg.state.vt.us/docs/2014/Acts/ACT120.pdf. Accessed on April 15, 2016.

Rule 702. Federal Rules of Evidence (2015)

The Federal Rules of Evidence were first adopted in 1975. They have been updated and revised a number of times since then. Rule 702 is now accepted in most courts as the standard for admissibility of scientific evidence in a trial, as explained earlier. The rule reads as follows:

A witness who is qualified as an expert by knowledge, skill, experience, training, or education may testify in the form of an opinion or otherwise if:

(a) the expert's scientific, technical, or other specialized knowledge will help the trier of fact to understand the evidence or to determine a fact in issue;

(b) the testimony is based on sufficient facts or data;

(c) the testimony is the product of reliable principles and methods; and

(d) the expert has reliably applied the principles and methods to the facts of the case.

Source: "Current Rules of Practice and Procedure." 2016. United States Courts. http://www.uscourts.gov/rules-poli cies/current-rules-practice-procedure. Accessed on April 16, 2016.

Grocery Manufacturers Association, et al. v. William H. Sorrell, et al. Case No. 5:14-cv-117 (2015)

As might be expected, the state of Vermont's decision to require the labeling of genetically modified foods in 2014 brought objections from a number of food manufacturing companies and related businesses about the constitutionality of such an act. A group of those companies filed suit against the state of Vermont, a case that was heard in the U.S. District Court for the District of Vermont in late January 2015. The case was somewhat complicated and involved five separate counts on which the judge, Christina Reiss, gave a mixed ruling, allowing some complaints, but not others. In effect, however, the law was allowed to stand and to take effect on the designated date of July 1, 2016. Counts One, Two, and Three all argued that the law interfered with the plaintiffs' freedom of speech under the First Amendment of the U.S. Constitution. Count Four argued that the law violated the U.S. Commerce Clause. And Count Five claimed that the law was superceded by federal statutes. Major elements of the judge's decision are as follows:

In Count One of their Amended Complaint, Plaintiffs challenge Act 120's GE disclosure requirement, alleging that:

Act 120 compels manufacturers to use labels that do not accurately describe their products, that could confuse consumers rather than inform them, and that could frighten consumers from purchasing safe, nutritious, affordable foods that are no different from counterpart organic, "Non-GMO" certified, or otherwise exempted foods. At bottom, Act 120 requires manufacturers to use their labels to convey an opinion with which they disagree, and that the State does not purport to endorse: namely, that consumers should assign significance to the fact that a product contains an ingredient derived from a genetically engineered plant. . . .

To the extent Count One of the Amended Complaint asserts claims that Act 120's GE disclosure requirement must be invalidated on the basis of strict scrutiny, those claims are not plausible and are hereby DISMISSED. . . .

In Count Two of the Amended Complaint, Plaintiffs assert a First Amendment challenge to Act 120's "natural" restriction, which prohibits GE manufacturers from using labeling, advertising, or signage indicating that a GE food product is "'natural,' 'naturally made,' 'naturally grown,' 'all natural,' or any words of similar import that would have a tendency to mislead a consumer." 9 V.S.A. § 3043(c). They allege that the State cannot establish that the restricted terms are inherently misleading, actually misleading, or potentially misleading when applied to GE foods and that, even if the State could make this showing, the restriction does not materially advance the State's asserted interests and is more extensive than necessary. The State seeks dismissal of Count Two, arguing that Plaintiffs fail to state a claim for which relief may be granted because GE manufacturers' use of "natural" terminology is entitled to no protection under the First Amendment. . . .

Plaintiffs have stated a plausible claim that Act 120's "natural" restriction is invalid under the First Amendment. They have further established that they are likely to succeed on the merits of this claim at trial. The State's motion to dismiss Count Two of the Amended Complaint is therefore DENIED. . . .

In Count Three, Plaintiffs contend that Act 120's restriction on "any words of similar import," 9 V.S.A. § 3043(c), is void-for-vagueness under the First Amendment and the Due Process Clause because it fails to provide reasonable notice of the scope of conduct that gives rise to civil penalties and authorizes arbitrary enforcement actions. As Act 120's "natural" restriction extends beyond food product labeling and covers advertising and signage activities as well, Plaintiffs ask that Act 120 be declared void in its entirety. . . .

The State's motion to dismiss Count Three for failure to state a claim under Fed. R. Civ. P. 12(b)(6) is DENIED. The court further finds that Plaintiffs are likely to succeed on the merits of their facial void-for-vagueness challenge to Act 120's prohibition on the use of "any words of similar import" at trial. . . .

In Count Four, Plaintiffs allege Act 120 violates the dormant Commerce Clause, and on that basis they ask that the Act be declared invalid in its entirety. . . .

[the court] DISMISSES Count Four with respect to Plaintiffs' claims that Act 120 violates the Commerce Clause, with the exception of Plaintiffs' per se challenge to the application of Act 120's "natural" restriction to GE manufacturers' nationwide and Internet signage and advertising activities; . . .

In Count Five of their Amended Complaint, Plaintiffs allege that Act 120 is expressly preempted or conflict preempted, in whole or in part, by various federal laws and thus violates Article VI, Clause 2 of the U.S. Constitution (the "Supremacy Clause"). The Supremacy Clause provides that the laws of the United States are "the supreme Law of the Land; . . . any Thing in the Constitution or Laws of any State to the Contrary notwithstanding." . . .

. . . the State's motion to dismiss Plaintiffs' Supremacy Clause claims in Count Five of the Amended Complaint is GRANTED IN PART and DENIED IN PART. Plaintiffs' Supremacy Clause claims alleging express and conflict preemption under the FDCA and-NLEA are DISMISSED, as well as their claim of conflict preemption pursuant to the Coordinated Framework. . . .

In conclusion: With regard to Plaintiffs' claims that have not been dismissed, the court DENIES Plaintiffs' motion for a preliminary injunction.

Source: Grocery Manufacturers Association, et al. v. William H. Sorrell, et al. Case No. 5:14-cv-117. U.S. District Court for the District of Vermont. http://www.centerforfoodsafety.org/files/vermont-decision_81793.pdf. Accessed on April 15, 2016.

People v. Collins; NY Slip Op 25227 [49 Misc 3d 595] (2015)

One of the most contentious issues in DNA analysis today, especially when applied to forensic cases, has to do with so-called low-copy number (LCN) DNA. The term refers to studies conducted on very small amounts of DNA, sometimes no more than a few dozen molecules of DNA, or even less. Questions have arisen as to the reliability of evidence adduced from LCN DNA because of the relatively high risk of contamination of samples being studied. As of 2016, many law enforcement agencies and courts have accepted the validity of LCN DNA evidence, although many have also declined to accept such evidences. In late 2015, acting Brooklyn Supreme Court justice Mark Dwyer issued a ruling in this case in which he declined to accept LCN DNA data for the reasons mentioned here. His reasoning was as follows (triple asterisks indicate omitted citations):

As a result of issues like these [discussed above], no public laboratory in the United States, other than the OCME [New York City Office of the Chief Medical Examiner] lab, employs high sensitivity analysis to develop profiles for use in criminal cases. Among the labs refusing to use high sensitivity analysis is the FBI laboratory. Moreover, CODIS, which is run by the FBI and contains the national DNA database, will not upload profiles created with high sensitivity analysis. Some laboratories, including Dr. Budowle's lab at the University of North Texas, will use high sensitivity analysis for limited purposes. For example, after a disastrous accident like an airplane crash, high sensitivity analysis of bodily remains can be used to identify the victims of the "closed" population of possible contributors. But that is because the population is limited, and because the remains, for example bones, can be cleaned before the analysis is done. And, unfortunately, a mistaken analysis can be of no consequence to the contributor.

Except for OCME, then, no American laboratory produces high sensitivity conclusions for use as evidence in a criminal case. As Dr. Budowle notes, that does not mean that high sensitivity analysis must be considered totally irrelevant in criminal cases. Such analysis can produce "investigative leads." Critics of high sensitivity analysis agree that if a DNA profile created through high sensitivity analysis suggests that a particular individual is the perpetrator of a crime, that profile can legitimately point investigators at the suspect. In that regard, the results of some other techniques—polygraphs and facial recognition software, for example—likewise can aid an investigation, but are not considered sufficiently reliable to be admissible at a trial.

*** This court initially wondered why the criticisms of high sensitivity analysis were not matters of weight to be considered by the jury—particularly since even defense witnesses like Dr. Budowle acknowledge that a profile produced by such analysis can be of value. Ultimately, however, that thought is trumped by Frye. The products of polygraph technology and of facial recognition technology similarly can sometimes have value, but evidence produced by those technologies is not generally accepted as reliable by the relevant scientific communities and so cannot be admitted in trials. The same should be true, at least at this time, for high sensitivity analysis. After all, if the experts in the DNA field cannot agree on the weight to be given to evidence produced by high sensitivity analysis, it would make no sense to throw such evidence before a lay jury and ask the jurors to give the evidence appropriate weight.

The People insist, however, that the relevant scientific community does accept high sensitivity analysis. It is true, as the People note, that OCME's procedures have been described in peer-reviewed articles and in discussions at gatherings of scientists. But this court cannot accept the thesis that publication and discussions equate to general acceptance. Not only the impressive defense witnesses indicate otherwise, so too do the many peer-reviewed articles submitted as defense exhibits which question OCME's procedures. And, as the defense notes, after all this discussion of high sensitivity analysis, no other laboratory has employed it for use in criminal cases. This court simply cannot conclude that there is a general consensus in favor of high sensitivity analysis, in the face of this contrary evidence.

The People have a more specific argument that decidedly deserves attention. High sensitivity analysis was approved by the DNA subcommittee of the New York State Forensic Commission. The conclusions of that subcommittee are binding on the Commission, and so the subcommittee numbers are the true decision-makers. And the members of the subcommittee are world-class scientists in various disciplines relevant to DNA analysis. The subcommittee approved OCME's high sensitivity procedures, and the People suggest that this is very strong evidence of general acceptance in the relevant scientific community.

This court does not agree. It is not just that Dr. Chakraborty, one of the members of the subcommittee, has "defected," and now has testified for the defense. The more important point is *** that no state subcommittee can be equated with the general membership of the relevant scientific community. Will we next consider the matter closed, because the members of a committee in Idaho or Florida approve of a procedure? This court knows that the members of the DNA subcommittee are indeed experts in their particular fields and that their opinions are valuable. They simply are not determinative.

Nor does the court agree with the People that the OCME validation studies and the audits at the OCME laboratory by outside reviewers are conclusive. Every laboratory validates techniques and procedures before implementing them. But a laboratory's satisfaction with its validation results does not show general acceptance of techniques and procedures, if the validation studies fail to create such general acceptance. And OCME's validation studies have failed to create general acceptance of high sensitivity analysis. As to audits, they appear to test whether procedures are being implemented in accordance with protocols, not whether the principles underlying the procedures are valid.

Source: *People v. Collins.* 2015. 2015 NY Slip Op 25227 [49 Misc 3d 595]. http://www.courts.state.ny.us/Reporter/3d series/2015/2015_25227.htm. Accessed on March 25, 2016.

Collection and Use of DNA Identification Information from Certain Federal Offenders 42 U.S. Code § 14135a (2016)

The Violent Crime Control and Law Enforcement Act of 1994 authorized the Federal Bureau of Investigation (FBI) to create a database of DNA samples taken from a select group of individuals who had committed certain types of federal crimes. The FBI did not interpret that act as authorizing its collection of DNA samples, and asked Congress to clarify that aspect of its database program. Congress resolved that issue by passing the DNA Analysis Backlog Elimination Act of 2000, and subsequent legislation, both authorizing the FBI to collect DNA samples and to gradually expand the category of individuals from whom samples could be taken. Today, the ultimate consequence of these laws is codified in section 14135A of Title 45 of the U.S. Code, which is reproduced here.

§14135a. Collection and Use of DNA Identification Information from Certain Federal Offenders

(a) Collection of DNA samples

(1) From Individuals in custody

(A) The Attorney General may, as prescribed by the Attorney General in regulation, collect DNA samples from individuals who are arrested, facing charges, or convicted or from non-United States persons who are detained under the authority of the United States. The Attorney General may delegate this function within the Department of Justice as provided in section 510 of title 28 and may also authorize and direct any other agency of the United States that arrests or detains individuals or supervises individuals facing charges to carry out any function and exercise any power of the Attorney General under this section.

(B) The Director of the Bureau of Prisons shall collect a DNA sample from each individual in the custody of the Bureau of Prisons who is, or has been, convicted of a qualifying Federal offense (as determined under subsection (d) of this section) or a qualifying military offense, as determined under section 1565 of title 10.

(2) From individuals on release, parole, or probation

The probation office responsible for the supervision under Federal law of an individual on probation, parole, or supervised release shall collect a DNA sample from each such individual who is, or has been, convicted of a qualifying Federal offense (as determined under subsection (d) of this section) or a qualifying military offense, as determined under section 1565 of title 10.

(3) Individuals already in CODIS

For each individual described in paragraph (1) or (2), if the Combined DNA Index System (in this section referred to as "CODIS") of the Federal Bureau of Investigation contains a DNA analysis with respect to that individual, or if a DNA sample has been collected from that individual under section 1565 of title 10, the Attorney General, the Director of the Bureau of Prisons, or the probation office responsible (as applicable) may (but need not) collect a DNA sample from that individual.

(4) Collection procedures

(A) The Attorney General, the Director of the Bureau of Prisons, or the probation office responsible (as applicable) may use or authorize the use of such means as are reasonably necessary to detain, restrain, and collect a DNA sample from an individual who refuses to cooperate in the collection of the sample.

(B) The Attorney General, the Director of the Bureau of Prisons, or the probation office, as appropriate, may enter into agreements with units of State or local government or with private entities to provide for the collection of the samples described in paragraph (1) or (2).

(5) Criminal penalty

An individual from whom the collection of a DNA sample is authorized under this subsection who fails to cooperate in the collection of that sample shall be—

(A) guilty of a class A misdemeanor; and

(B) punished in accordance with title 18.

(b) Analysis and use of samples

The Attorney General, the Director of the Bureau of Prisons, or the probation office responsible (as applicable) shall furnish each DNA sample collected under subsection (a) of this section to the Director of the Federal Bureau of Investigation, who shall carry out a DNA analysis on each such DNA sample and include the results in CODIS.

(c) Definitions

In this section:

(1) The term "DNA sample" means a tissue, fluid, or other bodily sample of an individual on which a DNA analysis can be carried out.

(2) The term "DNA analysis" means analysis of the deoxyribonucleic acid (DNA) identification information in a bodily sample.

(d) Qualifying Federal offenses

The offenses that shall be treated for purposes of this section as qualifying Federal offenses are the following offenses, as determined by the Attorney General:

(1) Any felony.

(2) Any offense under chapter 109A of title 18.

(3) Any crime of violence (as that term is defined in section 16 of title 18).

(4) Any attempt or conspiracy to commit any of the offenses in paragraphs (1) through (3).

Section (e) and (f) deal with administrative issues related to the law.

Source: §14135a. Collection and Use of DNA Identification Information from Certain Federal Offenders. 2016. United

States Code. Office of the Law Revision Counsel. http://uscode
.house.gov/view.xhtml?req=(title:42%20section:14135a%20
edition:prelim)%20OR%20(granuleid:USC-prelim-title
42-section14135a)&f=treesort&edition=prelim&num=0&jump
To=true. Accessed on April 17, 2016.

Issues of DNA Collection

*One of the fundamental issues that have long been raised with re-
gard to the development of DNA databases is the extent to which
law enforcement official ("the state") can impose on a person's privacy to collect DNA samples. Is such a process of sufficient importance to the state in controlling crime to intrude on a person's
privacy by taking a sample of his or her bodily fluids? As of 2016,
that issue is still a topic of debate within U.S. courts. In 2013,
the U.S. Supreme Court decided on a 5-to-4 vote that the state's
interests trumped privacy rights. At about the same time, however,
some state courts were taking the opposite view and prohibiting
law enforcement officials from requiring DNA samples from individuals who had not been arrested for, or found guilty of, certain
crimes. The short segments here provide the basic reasoning on both
sides of this issue.*

Maryland v. King 569 U.S. ____

A suspect's criminal history is a critical part of his identity
that officers should know when processing him for deten-
tion. . . . Police already seek this crucial identifying information. They use routine and accepted means as varied as
comparing the suspect's booking photograph to sketch art-
ists' depictions of persons of interest, showing his mugshot
to potential witnesses, and of course making a computerized
comparison of the arrestee's fingerprints against electronic
databases of known criminals and unsolved crimes. In this
respect the only difference between DNA analysis and the
accepted use of fingerprint databases is the unparalleled accuracy DNA provides. . . .

In sum, there can be little reason to question "the legitimate interest of the government in knowing for an absolute certainty the identity of the person arrested, in knowing whether he is wanted elsewhere, and in ensuring his identification in the event he flees prosecution." . . .

By comparison to this substantial government interest and the unique effectiveness of DNA identification, the intrusion of a cheek swab to obtain a DNA sample is a minimal one. True, a significant government interest does not alone suffice to justify a search. The government interest must outweigh the degree to which the search invades an individual's legitimate expectations of privacy. In considering those expectations in this case, however, the necessary predicate of a valid arrest for a serious offense is fundamental.

Source: Maryland v. King, 2013. Supreme Court of the United States. http://www.supremecourt.gov/opinions/12pdf/12-207_d18e.pdf. Accessed on April 17, 2016.

State v. Medina, et al. 2014 VT 69 (2014)

¶ 44. We find a broad warrantless-search authorization, under the theory that it is a search incident to an arrest, to be inconsistent with the requirements of Article 11 *[of the Vermont constitution]* as we have developed them. To the extent we have recognized the validity of a warrantless search incident to an arrest, it has been in cases where exigent circumstances were present. While it is possible that the fruits of a DNA search will produce information bearing on conditions of release or confinement with respect to a particular defendant, that possibility alone is insufficient to justify a warrantless DNA search of every defendant, with no distinction among those who will be searched.

¶ 45. In reaching this conclusion, we recognize that we have never held that a warrantless booking search of a detainee's person or property is inconsistent with Article 11 or that routine fingerprinting of arrestees is prohibited by Article 11.

However we decide the validity of these routine practices under Article 11, they do not justify the DNA sample capture involved here. We do not equate a procedure that takes a visible image of the surface of the skin of a finger with the capture of intimate bodily fluids, even if the method of doing so is speedy and painless. More important, despite the occasional usefulness of DNA samples for ordinary identification as described in King, the real functionality, and statutory purpose, is to solve open criminal cases or ones that may occur in the future. While part of this functionality may respond to a special need as we held in Martin, it is far afield from the immediate concern for the protection of arresting officers or the destruction of evidence, the concerns underlying our search-incident-to-arrest doctrine. The real expansion of warrantless search power in King is "its reimagination of the idea of 'identity' to include criminal history and other information." . . . Despite the assurances of the Court in King, it is difficult to see any limit on what information may be gathered about an arrestee and the effect of that information gathering on the decision whether to arrest.

. . .

Because of the limited weight of the State's interest in the expansion of the DNA sampling requirement to defendants on arraignment for a qualifying crime, and the greater privacy interest of the defendant at that stage of the adjudication, we—like the Minnesota Court of Appeals in Welfare of C.T.L.—conclude that the balance tips to the defendant. We also concur in the analysis of the Arizona Supreme Court that "[h]aving a DNA profile before adjudication may conceivably speed . . . investigations [of other crimes]. . . . But one accused of a crime, although having diminished expectations of privacy in some respects, does not forfeit [constitutional] protections with respect to other offenses not charged absent either probable cause or reasonable suspicion." . . .

¶ 63. The marginal weight of the State's interest in DNA collection at the point of arraignment, balanced against the weight of the privacy interest retained by arraignees prior to conviction, persuades us to hold that 20 V.S.A. § 1933(a)(2), and associated sections, which expand the DNA-sample requirement to defendants charged with qualifying crimes for which probable cause is found, violates Chapter I, Article 11 of the Vermont Constitution.

Source: State v. Medina, et al., 2014. Vermont Supreme Court. https://outside.vermont.gov/dept/VTLIB/Documents/Medina.pdf. Accessed on April 17, 2016.

The term *DNA technology* covers a very wide range of topics and is, therefore, discussed in an extensive collection of books, articles, reports, and Internet pages. This chapter can provide information on only a small number of those references. Readers should also refer to the references listed at the end of Chapters 1 and 2 to complement this list.

Many articles now appear in both print and electronic forms. For items of that type listed in this chapter, both formats are listed.

Books

Agus, David, and Kristin Loberg. 2016. *The Lucky Years: How to Thrive in the Brave New World of Health.* New York: Simon & Schuster.

> The authors review some of the important breakthroughs in the health sciences resulting from new knowledge about DNA technology, with some added suggestions for dealing intelligently with these new health strategies.

Al-Bar, Mohammed Ali, and Hassan Chamsi-Pasha. 2015. "Ethical Issues in Genetics (Premarital Counseling, Genetic

---

A customer reviewing a DNA self-testing kit that can be used to screen for 17 different diseases and conditions. A review of the test costs a customer $2,500. (Shaul Schwarz/Getty Images)

Testing, Genetic Engineering, Cloning and Stem Cell Therapy, DNA Fingerprinting).” In Al-Bar, Mohammed Ali, and Hassan Chamsi-Pasha. *Contemporary Bioethics. Islamic Perspective*, Chapter 12. Cham, Switzerland: Springer.

    The authors provide views from the Islamic religion about a range of topics involving DNA technology, such as those listed in the chapter title.

Anthes, Emily. 2014. *Frankenstein's Cat: Cuddling up to Biotech's Brave New Beasts*. New York: Scientific American/Farrar, Straus and Giroux.

    The author describes in a reader-friendly way some of the transgenic animals that have been or might be created through the process of recombinant DNA technology.

Berliner, Janice L., ed. 2015. *Ethical Dilemmas in Genetics and Genetic Counseling: Principles through Case Scenarios*. Oxford, UK; New York: Oxford University Press.

    This collection of essays deals with a variety of topics in the field of current genetics, such as prenatal testing, testing for assisted reproductive technologies (ARTs), testing children for adult-onset disorders, and problems associated with incidental findings.

Boniolo, Giovanni, and Virginia Sanchini. 2016. *Counselling and Medical Decision-Making in the Era of Personalised Medicine: A Practice-Oriented Guide*. Publisher: [n.p.]: Springer.

    Designed for specialists in the field, this book provides a good deal of valuable information for the general reader, including a discussion of topics such as informed consent, ethical issues for patients and practitioners, the significance of probability in interpreting results, and incidental findings from testing.

Bradley, James T. 2013. *Brutes or Angels: Human Possibility in the Age of Biotechnology.* Tuscaloosa: The University of Alabama Press.
The author provides a good general introduction to the types of genetic engineering that are currently taking place in research laboratories and speculates about possible changes in human characteristics in the future. He then outlines some of the political, social, and other issues that may arise as a result of these potential changes.

Buckleton, John S., Jo-Anne Bright, and Duncan Taylor. 2016. *Forensic DNA Evidence Interpretation*, 2nd ed. Boca Raton, FL: CRC Press.
This book begins with a review of the chemistry and biology of DNA typing and then focuses on the statistical interpretation of data obtained from the analysis of DNA typing. The presentation tends to be quite technical, but it is of value even to the general reader.

Cathomen, Toni, Matthew Hirsch, and Matthew Porteus, eds. 2016. *Genome Editing: The Next Step in Gene Therapy.* New York: Springer.
The selection of articles in this anthology provide a review of the history of new genome-editing technology, along with a discussion of current research and potentials for the future of gene therapy.

Claiborne, Anne Ba., Rebecca A. English, and Jeffrey P. Kahn. 2016. *Mitochondrial Replacement Techniques: Ethical, Social, and Policy Considerations.* Committee on the Ethical and Social Policy Considerations of Novel Techniques for Prevention of Maternal Transmission of Mitochondrial DNA Diseases. Washington, DC: National Academies Press. Available online at http://www.nap.edu/read/21871/chapter/1. Accessed on April 22, 2016.
Mitochondrial replacement is a cloning technology for the treatment of mitochondrial disease, which prevents

women from completing a pregnancy successfully. The technology now appears to be technically feasible, although it has raised very significant ethical, moral, social, and other issues.

Clark, David P., and Nanette Jean Pazdernik. 2016. *Biotechnology: Applying the Genetic Revolution*, 2nd ed. Amsterdam; Boston: Academic Press/Elsevier.

This book provides a broad, general introduction to the subject of biotechnology and its many applications. Specific chapters deal with topics such as the basics of biotechnology; DNA, RNA, and protein; genomics and gene expression; protein engineering; transgenic plants and plant biotechnology; transgenic animals; gene therapy; biowarfare and bioterrorism; and ethical issues in the use of biotechnology.

Croce, Nicholas, 2016. *The Science and Technology behind the Human Genome Project*. New York: Britannica Educational Publishing.

This book provides a good general introduction to DNA technology written for students in high school.

Dale, Jeremy, Malcolm von Schantz, and Nick Plant. *From Genes to Genomes: Concepts and Applications of DNA Technology*, 3rd ed. Chichester, UK; Hoboken, NJ: John Wiley & Sons.

This book covers virtually every aspect of DNA technology, from an explanation of the molecular structure of DNA to the techniques used in DNA technology to the applications of DNA technology in today's world.

Doudna, Jennifer A., and Erik J. Sontheimer, eds. 2014. *The Use of CRISPR/Cas9, ZFNs, and TALENs in Generating Site-Specific Genome Alterations*. Amsterdam: Elsevier/Academic Press.

This highly technical book is intended for specialists in the field, but it also provides an invaluable introduction to one of the most important new fields in the science of DNA technology.

Dowell, David R. 2015. *Nextgen Genealogy: The DNA Connection*. Santa Barbara, CA: Libraries Unlimited
    The author explains how DNA analysis can be used for gaining knowledge about genealogy and what the future of this field of study is likely to look like.

Dreye, Malte, Jeanette Erdmann, and Christoph Rehmann-Sutter, eds. 2016. *Genetic Transparency?: Ethical and Social Implications of Next Generation Human Genomics and Genetic Medicine*. Leiden, the Netherlands: Brill Rodopi.
    The essays in this volume proceed from the expanded range of options for changing a person's genomic characteristics, raising questions as to what information should become available to an individual and to other individuals and organizations (if any).

Fletcher, Amy Lynn. 2014. *Mendel's Ark: Biotechnology and the Future of Extinction*. New York: Springer.
    The author explores some of the possibilities for saving endangered species and/or restoring recently or more distantly extinct species using modern methods of cloning.

Hesse-Biber, Sharlene Nagy. 2014. *Waiting for Cancer to Come: Women's Experiences with Genetic Testing and Medical Decision Making for Breast and Ovarian Cancer*. Ann Arbor: The University of Michigan Press.
    The author begins with a chapter discussing the genetic testing industry's motivations in promoting its products. The remainder of the book comments on preparation for the BRCA breast cancer test, its possible outcomes, and ways that women have developed for dealing with whatever news they may have received as a result of the test.

Hodge, Russ. 2009. *Genetic Engineering: Manipulating the Mechanisms of Life*. New York: Facts On File.

This book provides a general overview of the history, technology, and ethical issues involved in DNA technology. Various chapters focus on the principles of classical genetics and molecular genetics, the rise of recombinant DNA technology, some practical applications of rDNA technology, and ethical issues associated with genetic engineering.

Jamieson, Allan, and Scott Bader, eds. 2016. *A Guide to Forensic DNA Profiling*. Chichester, UK: John Wiley & Sons.

This book provides an excellent overview of essentially all the aspects of DNA typing in forensics, including collection of DNA samples, analysis and interpretation of samples, applications of forensic DNA typing, and issues involved in court proceedings.

Jerslid, Paul T. 2009. *The Nature of Our Humanity: Ethical Issues in Genetics and Biotechnology*. Minneapolis, MN: Fortress Press.

The author is professor emeritus of theology and ethics at Lutheran Theological Southern Seminary and outlines the theological and ethical issues that arise out of biotechnological research on humans in the field of medicine and scientific research.

Klitzman, Robert. 2012. *Am I My Genes?: Confronting Fate and Family Secrets in the Age of Genetic Testing*. Oxford; New York: Oxford University Press.

The author notes that advances in genetic testing have provided a mechanism by which individuals can learn a great deal more about their genetic history and the health issues that may raise. He describes the nature of genetic testing, information that can be gained from the process, and how individuals can process and make use of that information.

Lawson, Chalres, and Berris Charnley, eds. 2015. *Intellectual Property and Genetically Modified Organisms: A Convergence in Laws*, Farnham, Surrey, UK; Burlington, VT: Ashgate.

This collection of essays considers in some detail the legal issues related to the invention and application of genetically modified organisms.

Lemmens, Trudo, Mireille Lacroix, and Roxanne Mykitiuk. 2007. *Reading the Future?: Legal and Ethical Challenges of Predictive Genetic Testing.* Montréal: Éditions Thémis.

The authors examine a number of social, ethical, legal, and other issues raised by the increasing availability of a variety of genetic tests. They consider issues such as the need for genetic testing, regulations that may be necessary or appropriate, stigmatization that may result from certain types of genetic testing results, patenting of genes, commercialization of testing technologies, and clinical issues.

MacKellar, Calum, and Christopher Bechtel, eds. 2014. *The Ethics of the New Eugenics.* New York: Berghahn Books.

The papers presented in this anthology are based on the assumption that trends in modern genetics research mirror the eugenics movement of the 20th century. They explore the nature of that movement, the similarities between it and modern genetics research, and arguments for and against proceeding with this research.

Makin, David A. 2015. *DNA and Property Crime Scene Investigation: Forensic Evidence and Law Enforcement.* London: Routledge.

The premise of this book is that DNA analysis can be a powerful tool for the investigation of property crimes, just as it is in crimes against persons. In the process of discussing this point of view, the author provides an excellent and readily understandable review of the basic principles of DNA forensic technology.

McCabe, Lina L., and Edward R. B. McCabe. 2008. *DNA: Promise and Peril*. Berkeley: University of California Press.

This book presents a number of issues that have arisen and may arise from scientists' ability to manipulate the DNA structure of individual organisms. Topics include the use of DNA technology in forensic science, protection against genetic discrimination, the ownership of genes by corporations, reproductive technologies, therapeutic and reproductive cloning, genomic medicine, and gene therapy.

Minor, Jessia. 2015. *Informed Consent in Predictive Genetic Testing*. Cham, Switzerland: Springer.

The author argues that advances in genetic testing have raised new issues about the concept of informed consent. The book focuses on the changing field of predictive genetic testing, the history of informed consent, and a new model of informed consent that may be needed for the new field of genetic testing.

Newton, David E. 2008. *DNA Evidence and Forensic Science*. New York: Facts on File, 2008.

A book in the *Library in a Book* series that provides an extended introduction to the topic, along with a variety of resources to aid the average researcher, including a group of biographical sketches, a chronology, list of organizations, list of print and electronic resources, and glossary of important terms.

Newton, David E. 2015. *Cloning: A Reference Handbook*. Santa Barbara, CA: ABC-CLIO.

This book provides not only a general introduction to the subject of cloning, but also an array of resources that readers can use in learning more about the topic and/or their own research on the subject.

Oksanen, Markku, and Helena Siipi, eds. 2014. *The Ethics of Animal Re-creation and Modification: Reviving, Rewilding, Restoring*. Basingstoke, UK; New York: Palgrave Macmillan.

The papers in this book review the technical possibilities for cloning extinct species and, primarily, the ethical and philosophical questions raised by the use of such a technology.

Panno, Joseph. 2008. *Gene Therapy: Treating Disease by Repairing Genes*. New York: Facts on File.

This book is part of Fact on File's *New Biology* series, designed for high school students and general readers. It provides a review of the methodology and applications of gene therapy along with a number of research aids, including a chronology, set of biographical sketches, list of print and electronic resources, glossary, and list of organizations involved in gene therapy.

Parsons, Thomas J., and Victor Walter Weedn. 2006. "Preservation and Recovery of DNA in Postmortem Specimens and Trace Samples." In Haglund, William D., and Marcella H. Sorg, eds. *Forensic Taphonomy: The Postmortem Fate of Human Remains*, 9th ed., Chapter 7. Boca Raton, FL: CRC Press.

This chapter is devoted to a discussion of the ways in which DNA fingerprinting can be used to identify human remains after death, but prior to its discovery as a fossil.

Qaim, Matin. 2016. *Genetically Modified Crops and Agricultural Development*. Houndmills, Basingstoke, UK; New York: Palgrave Macmillan.

This book provides a general introduction to the technology of genetically modified crops, along with a review of potential benefits and risks associated with GM crops, new and future applications of the technology, public opinions, and appropriate forms of regulation.

Robinson, Claire, Michael Antoniou, and John Fagan. 2015. *GMO Myths and Truths: A Citizen's Guide to the Evidence on the Safety and Efficacy of Genetically Modified Crops*. London: Earth Open Source.

The authors attempt to make a strong case against the development and use of GM crops, arguing that the scientific evidence suggests that traditional non-GM approaches result in crops that are superior in many respects to GM crops.

Shapiro, Beth Alison. 2015. *How to Clone a Mammoth: The Science of De-Extinction*. Princeton, NJ: Princeton University Press.

This book discusses the way in which extinct species can be restored by means of cloning. Although the emphasis is on the science and technology of the process, it is written so as to be easily accessible to the general reader. The last chapter of the book talks about whether or not the procedure should actually be conducted.

Watson, Ronald R., and Victor R. Preedy, eds. 2016. *Genetically Modified Organisms in Food: Production, Safety, Regulation and Public Health*. Amsterdam: Elsevier Science.

This collection of essays focuses on a variety of technical topics related to the process of genetic engineering of foods, along with extensive discussions of a variety of social, political, economic, and legal issues related to the production and sale of these materials.

Woestendiek, John. 2010. *Dog, Inc.: The Uncanny Inside Story of Cloning Man's Best Friend*. New York: Avery.

The author provides a history of attempts to clone a variety of animals, including the cloning of pets, along with a description of the procedure and a review of the current state of research in the field.

Wolfe, James. 2016. *Genetic Testing and Gene Therapy.* New York: Britannica Educational Publishing.
This book is intended for young adults. It provides an excellent and up-to-date general introduction to the topics of genetic testing and gene therapy.

Articles

Aronson, Jay D. 2008. "Creating the Network and the Actors: The FBI's Role in the Standardization of Forensic DNA Profiling." *Biosocieties.* 3: 195–215.
The author provides an interesting historical review of the early development of DNA databases, issues raised about the development of this technology, and solutions developed by the FBI for dealing with the new technology of DNA fingerprinting.

Atala, Anthony. 2012. "Tissue Engineering of Reproductive Tissues and Organs." *Sterility and Fertility.* 98(1): 21–29.
This paper outlines the status of research on problems involving the human reproductive system resulting from injuries, old age, or other factors. The author notes that advances in stem cell research and therapeutic cloning may provide solutions for some of these problems.

Ayala, Francisco J. 2015. "Cloning Humans? Biological, Ethical, and Social Considerations." *Proceedings of the National Academy of Sciences of the United States of America.* 112(29): 8879–8886.
This paper was written by an eminent biologist and theologian. It covers a wide range of topics related to the modification of organisms, especially humans, by a variety of new genetic technologies. An excellent introduction to and overview of such issues.

Azadi, Hossein, et al. 2016. "Genetically Modified Crops and Small-Scale Farmers: Main Opportunities and Challenges." *Critical Reviews in Biotechnology.* 36(3): 434–446.

GM crops would appear to have a number of benefits for farmers of all sizes. Adoption of such crops has been minimal among smaller farmers, however. This article considers some of the reasons for this low level of adoption, along with some of the advantages that such crops may have for small as well as large farming units.

Bell, Karen L., et al. 2016. "Review and Future Prospects for DNA Barcoding Methods in Forensic Palynology." *Forensic Science International: Genetics.* 21: 110–116.

The authors discuss the use of DNA barcoding in the identification of pollen. They begin with a review of the history of forensic palynology (the study of pollen) and of DNA barcoding, with a consideration of the technical issues in the application of the latter to the former.

Berg, Cheryl, and Kelly Fryer-Edwards. 2008. "The Ethical Challenges of Direct-to-Consumer Genetic Testing." *Journal of Business Ethics.* 77(1): 17–31.

In 2005, 13 Web sites offered genetic testing kits for sale directly to consumers. This article examines the scientific, legal, and ethical implications of such offers in light of the fact that some kits used questionable and/or unverified scientific procedures, and little and/or inadequate information was provided on the use and interpretation of the tests.

Berg, Paul, and Janeet E. Mertz. 2010. "Personal Reflections on the Origins and Emergence of Recombinant DNA Technology." *Genetics.* 184(1): 9–17. Available online at http://www.ncbi.nlm.nih.gov/pmc/articles/PMC2815933/. Accessed on April 23, 2016.

One of the pioneers in the field of DNA technology writes about his personal remembrances of the early days of research in the field.

Bonny, Sylvie. 2016. "Genetically Modified Herbicide-Tolerant Crops, Weeds, and Herbicides: Overview and Impact." *Environmental Management*. 57(1): 31–48.

   The author considers the effect of GM crops on the development of resistant weeds in the United States and other parts of the world.

Brookes, Graham, and Peter Barfoot. 2014. "Economic Impact of GM Crops." *GM Crops & Food*. 5(1): 65–75.

   The authors attempt to estimate the economic benefits to agriculture that have resulted from the expanded use of GM crops around the world. They estimate that net economic benefits at the farm level amounted to $18.8 billion in 2012 and $116.6 billion over the 17-year period from 1996 to 2012.

Chao, Jie, et al. 2016. "DNA Nanotechnology-Enabled Biosensors." *Biosensors & Bioelectronics*. 76: 68–79.

   The authors describe research in which segments of DNA molecules are used as sensors for the detection of certain types of protein, nucleic acid, and other molecular targets.

Chambers, Geoffrey K., et al. 2014. "DNA Fingerprinting in Zoology: Past, Present, Future." *Investigative Genetics*. doi: 10.1 186/2041-2223-5-3. http://investigativegenetics.biomedcentral .com/articles/10.1186/2041-2223-5-3. Accessed on April 22, 2016.

   The authors begin with a discussion of the history of DNA fingerprinting as a forensic tool before describing its application to the field of experimental zoology. They suggest that the collection of large databases of DNA specimens from many specimens is a crucial key to the success of future studies.

Cooper, David K. C. 2015. "The Case for Xenotransplantation." *Clinical Transplantation*. 29(4): 288–293.

   The author, an authority in the field of xenotransplantation, argues that the technology is sufficiently developed to encourage more rapid research in its use for the supply of organs to humans.

Cooper, David K. C., et al. 2016. "The Role of Genetically Engineered Pigs in Xenotransplantation Research." *The Journal of Pathology*. 238(2): 288–299.

The authors describe some of the modifications that have been made in pigs to improve the likelihood of successful xenotransplantation in the near future, a prospect whose promise they outline in this article.

Coupe, Richard H., and Paul D. Capel. 2016. "Trends in Pesticide Use on Soybean, Corn and Cotton since the Introduction of Major Genetically Modified Crops in the United States." *Pest Management Science*. 72(5): 1013–1022.

One supposed advantage of the use of GM crops is that less pesticide may be required on lands where such crops are grown. The authors explore that argument and find that the use of herbicides on GM corn crops has been reduced, on cotton crops it has remained constant, and on soybeans it has increased. They explore possible reasons for these trends.

Cunningham, Thomas V. 2013. "What Justifies the United States Ban on Federal Funding for Nonreproductive Cloning?" *Medicine, Health Care and Philosophy*. 16(4): 825–841.

The author analyzes the justifications that have been provided for the ban on the use of public funds for research on therapeutic cloning and offers an argument as to why that ban are wrong.

Delaunay, Catarina. 2015. "The Beginning of Human Life at the Laboratory: The Challenges of a Technological Future for Human Reproduction." *Technology in Society*. 40: 14–24.

The author notes the increasing tendency to separate the process of reproduction from actual sexual acts and asks what social issues are likely to arise as a result of this trend in the future. She discusses some of the consequences of those new social problems.

Ethics Committee of the American Society for Reproductive Medicine. 2016. "Human Somatic Cell Nuclear Transfer and Reproductive Cloning: An Ethics Committee Opinion." *Fertility and Sterility.* 105(4): e1–e4.

The Ethics Committee of the American Society for Reproductive Medicine argues against the use of SCNT for human reproductive cloning "due to concerns about safety; the unknown impact of SCNT on children, families, and society; and the availability of other ethically acceptable means of assisted reproduction."

Evans, Barbara J., Wylie Burke, and Gail P. Jarvik. 2015. "The FDA and Genomic Tests—Getting Regulation Right." *New England Journal of Medicine.* 372(23): 2258–2264.

The authors provide an extensive review of recent actions by the U.S. Food and Drug Administration associated with the regulation of gene testing and the effect of such testing on further research and implementation of such testing practices in the United States.

Fadel, Hossam E. 2012. "Developments in Stem Cell Research and Th erapeutic Cloning: Islamic Ethical Positions, a Review." *Bioethics.* 26(3): 128–135.

The author discusses the ethical issues associated with stem cell research and therapeutic cloning in general, and reviews the status of such research in Muslim countries.

Ferrer-Miralles, Neus, et al. 2016. "Recombinant Pharmaceuticals from Microbial Cells: A 2015 Update." *Microbial Cell Factories.* 15(1): 1–7.

This article provides information on the current status of protein production by recombinant DNA methods, with data and statistics on the present state of the industry.

Gerasimova, Ksenia. 2016. "Debates on Genetically Modified Crops in the Context of Sustainable Development." *Science and Engineering Ethics.* 22(2): 525–547.

    The author discusses some issues in the debate over the development and use of genetically modified crops, especially in the areas of environmental effects, social and economic development, loss of biodiversity, and food security.

Hasson, Sidgi Syed Anwer Abdo, Juma Khalifa Zayid Al-Busaidi, and Talal Abdulmalek Sallam. 2015. "The Past, Current, and Future Trends in DNA Vaccine Immunisations." *Asian Pacific Journal of Tropical Biomedicine.* 5(5): 344–353. Available online at http://www.sciencedirect.com/science/article/pii/S2221169115 30366X. Accessed on April 23, 2016.

    The authors provide a detailed and comprehensive review of the history of DNA vaccine research, along with some suggestions as to its future role in preventative medicine.

Hemphill, Thomas A., and Syagnik Banerjee. 2015. "Genetically Modified Organisms and the U.S. Retail Food Labeling Controversy: Consumer Perceptions, Regulation, and Public Policy." *Business and Society Review.* 120(3): 435–464.

    This article provides a very thorough review of all aspects of the controversy over the labeling of GM foods, including the arguments pro and con for the practice, state and federal laws with regard to labeling, public opinion about labeling in the United States, current scientific evidence on the safety of GM foods, and legal actions related to the labeling of GM foods.

Hess, Pascale G., H. Mabel Preloran, and C. H. Browner. 2009. "Diagnostic Genetic Testing for a Fatal Illness: The Experience of Patients with Movement Disorders." *New Genetics and Society.* 28(1): 3–18.

    One of the crucial issues involved in the use of genetic testing is how patients make use of information obtained from the procedure. That issue is especially relevant when the disease being diagnosed is life-threatening. The authors of this article report on a study of 27 neurological consultations and 27 in-depth interviews with patients diagnosed with a fatal disease as a result of genetic testing.

Holtz, Barry R., et al. 2015. "Commercial-Scale Biotherapeutics Manufacturing Facility for Plant-Made Pharmaceuticals." *Plant Biotechnology Journal.* 13(8): 1180–1190.

This article describes the production of a large manufacturing facility capable of producing 350 kilograms of plant biomass for the production of an H1N1 vaccine as an example of the advanced state of pharmaceutical production using engineered plants.

Huesing, Joseph E., et al. 2016. "Global Adoption of Genetically Modified (GM) Crops: Challenges for the Public Sector." *Journal of Agricultural and Food Chemistry.* 64(2): 394–402.

Many experts believe that genetically modified crops hold great promise for relieving some of the world's most severe food shortages. The 2014 IUPAC International Congress of Pesticide Chemistry held in San Francisco included a symposium on "Challenges Associated with Global Adoption of Agricultural Biotechnology" to review current obstacles in promoting GM crops. This article provides a summary of the points made at that symposium.

Jeffreys, Alec J., Victoria Wilson, and Swee Lay Thein. 1985. "Hypervariable 'Minisatellite' Regions in Human DNA." *Nature.* 314(6006): 67–73.

Although very technical, this article is of historical importance as the first scholarly article describing the process of DNA testing as a means of identifying individual humans.

Jhansi Rani, S., and R. Usha. 2013. "Transgenic Plants: Types, Benefits, Public Concerns and Future." *Journal of Pharmacy Research.* 6(8): 879–883.

The authors provide a good general introduction to the topic of transgenic plants, with a brief discussion of the methods by which such plants are produced, a number of specific applications, and some possible issues for the technology in the future.

Kim, M.J., et al. 2012. "Lessons Learned from Cloning Dogs." *Reproduction in Domestic Animals.* 47(Suppl. 2): 115–119.

The authors review research that has been conducted at Seoul National University on the cloning of dogs, with information gained from this research for future applications.

Kim, Pil Ho, et al. 2015. "The Application of DNA Chip Technology to Identify Herbal Medicines: An Example from the Family Umbelliferae." *Natural Product Sciences.* 21(3): 185–191.

For consumers of herbal medicines, knowing that a product is what it is claimed to be can be an important issue. Since many herbal products are made available in forms (dried products, powders, etc.) that make it impossible for them to be identified by morphological criteria, other methods of identification may be necessary. This article describes the process by which a specially designed DNA chip can provide that means of identification.

Krause, Kenneth W. 2012. "New Life for Human (Therapeutic) Cloning? Whence Come the Eggs?" *Skeptical Inquirer.* 36(2): 28–30.

The author discusses the current status of therapeutic cloning, some of the technical issues involved, and how those issues affect the development of the technology.

Krimsky, Sheldon. "From Asilomar to Industrial Biotechnology: Risks, Reductionism and Regulation." *Science as Culture.* 14(4): 309–323.

Recombinant DNA research has always been beleaguered by ethical, social, economic, political, and legal issues. This essay reviews the evolution of those issues beginning with the Asilomar Conference held in February 1975 when the risks of rDNA were first discussed and acted upon by a group of scientists in the field.

Lanza, Robert P., et al. 2004. "Cloning of an Endangered Species (*Bos gaurus*) Using Interspecies Nuclear Transfer." *Cloning*. 2(2): 79–90.
This paper reports on the cloning of the first endangered species, a gaur, who survived less than 48 hours after its birth. Its death occurred because of a common infection among cattle and not as a result of the cloning experiment.

Lucht, Jan M. 2015. "Public Acceptance of Plant Biotechnology and GM Crops." *Viruses*. 7(8): 4254–4281.
This review article discusses the very wide differences of opinion about GM crops among the general public and scientists. Special attention is given to current debates over labeling of GM foods in the United States, a "hardening" of opinion in Europe against the use of GM foods, and the evolving role of GM foods in China.

May, Joshua. 2016. "Emotional Reactions to Human Reproductive Cloning." *Journal of Medical Ethics*. 42(1): 26–30.
The author points out that most surveys about human reproductive cloning ask about respondent's moral feelings toward the practice, not their emotional responses to it. This survey deals with the latter question.

Nicolás, Pilar. 2009. "Ethical and Juridical Issues of Genetic Testing: A Review of the International Regulation." *Critical Reviews in Oncology/Hematology*. 69(2): 98–107.
The author begins with a review of the international legal framework for regulations dealing with genetic testing, primarily the Convention on Human Rights and Biomedicine

and the International Declaration on Human Genetic Data, before discussing the objectives of providing protection for patients and the rights of patients involved in genetic testing procedures.

Niemann, Heiner, and Andrea Lucas-Hahn. 2012. "Somatic Cell Nuclear Transfer Cloning: Practical Applications and Current Legislation." *Reproduction in Domestic Animals.* 47(Suppl 5): 2–10.

This article provides a general introduction to the use of SCNT in the cloning of domestic animals and reviews the current legislative state of affairs in the European Union.

Ousterout, D.G., and C.A. Gersbach. 2016. "The Development of TALE Nucleases for Biotechnology." *Methods in Molecular Biology.* 1338: 27–42.

This article provides a general introduction to the field of TALE nuclease methods in gene editing, including a discussion of the history of their development, the technology involved, and some specific examples of their use in research.

Parisi, Claudia, Pascal Tillie, and Emilio Rodríguez-Cerezo. 2016. "The Global Pipeline of GM Crops Out to 2020." *Nature Biotechnology.* 34(1): 31–36.

The authors review current developments in GM crops and attempt to predict changes that are likely to occur over the next five years.

Roberts, David J. 2015. "A Kodak Moment for Law Enforcement: Using DNA Blueprints to Build Facial Composites." *The Police Chief.* 82(3): 78–79. Available online at http://www.poli cechiefmagazine.org/magazine/index.cfm?fuseaction=display_ arch&article_id=3667&issue_id=32015. Accessed on April 20, 2016.

DNA data can now be used to produce visual representations of the individuals from which the DNA was taken to the level of accuracy that allows the photograph to be used to identify possible suspects in crime investigations.

Schiml, Simon, and Holger Puchta. 2016. "Revolutionizing Plant Biology: Multiple Ways of Genome Engineering by CRISPR/Cas." *Plant Methods.* 12: 8. doi: 10.1186/s13007-016-0103-0.
   This article provides a general overview of the way in which CRIPS-Cas9 technology is being used in plant modification researches, as well as some of the various methods that have been designed for the new technology.

Siegel, Bernard, and Arnold I. Friede. 2013. "The U.S. Food and Drug Administration Should Solidify the Legal Basis for Its Authority over Reproductive Cloning." *Stem Cells and Development.* 22(Suppl 1): 46–49.
   The authors discuss problems of confusion between reproductive and therapeutic cloning and recommend that the FDA take a more assertive stance as to its regulatory position on the issue.

"The Threat of Human Cloning: Ethics, Recent Developments, and the Case for Action." 2015. *The New Atlantis.* 46: 9–146. Available online at http://www.thenewatlantis.com/docLib/20150825_TNA46TheThreatofHumanCloning.pdf. Accessed on April 22, 2016.
   The entire volume of this issue is devoted to a report by the Witherspoon Council on Ethics and the Integrity of Science. Various chapters of the report deal with the scientific and historical background of cloning, the case against both reproductive and therapeutic cloning, cloning policy in the United States, and a detailed summary of states laws on all forms of cloning.

Ugochukwu, Albert I., et al. 2015. "An Economic Analysis of Private Incentives to Adopt DNA Barcoding Technology for Fish Species Authentication in Canada." *Genome.* 58(12): 559–567. Available online at https://tspace.library.utoronto .ca/bitstream/1807/70304/1/gen-2015-0033.pdf. Accessed on April 20, 2016.

> The authors note that the substitution of fish in the economic chain from sources to the market have significant economic impact for producers and consumers. They suggest that the use of DNA barcoding would have very significant benefits for consumers in providing authentication of fish species offered for sale.

Wolt, Jeffrey D., Kan Want, and Bing Yang. 2016. "The Regulatory Status of Genome-Edited Crops." *Plant Biotechnology Journal.* 14(2): 510–518.

> The production of genetically engineered plants with new genome-editing tools raises new regulatory issues, since such plants may be identical to plants developed by non-engineered procedures that are not covered by current regulations. The authors suggest some issues that have been raised and possible ways of dealing with these issues.

Yi, Doogab. "Cancer, Viruses, and Mass Migration: Paul Berg's Venture into Eukaryotic Biology and the Advent of Recombinant DNA Research and Technology, 1967–1980." *Journal of the History of Biology.* 41(4): 589–636.

> An interesting survey of the early history of recombinant DNA research and a significant change in the paradigm of that research based on the work of Paul Berg. The latter portions of the article discuss some early and more recent applications of recombinant DNA research.

Reports

Connors, Edward, et al. 1996. "Convicted by Juries, Exonerated by Science: Case Studies in the Use of DNA Evidence to

Establish Innocence After Trial." Washington, DC: Office of Justice Programs. U.S. Department of Justice.

This report reviews the use of DNA typing in the United States to exonerate men and women convicted of crimes, usually rape, aggravated assault, and/or murder. The report offers a number of recommendations to increase the efficacy of DNA typing in proving the innocence of prisoners and concludes with detailed reviews of more than two dozen individuals whose sentences were commuted as a result of DNA testing.

Council of Europe. 2008. "Additional Protocol to the Convention on Human Rights and Biomedicine, Concerning Genetic Testing for Health Purposes." Strasbourg: Council of Europe.

In 1997, the Council of Europe adopted the Convention on Human Rights and Biomedicine, which was later amended four times. The last of these amendments, this document, was adopted after a long discussion of issues related to developments in modern biotechnology as they relate to genetics and human health. So many issues were involved in this discussion that members of the group working on this topic decided to limit their discussions to the topic of genetic testing for health purposes. The final conclusion of those discussions was this document, which is also available online at https://rm.coe.int/CoERM PublicCommonSearchServices/DisplayDCTMContent? documentId=0900001680084824.

Cowan, Tadlock, and Geoffrey S. Becker. 2009. "Biotechnology in Animal Agriculture: Status and Current Issues." Washington, DC: Congressional Research Service.

This report reviews the current status of recombinant DNA research on animals, regulation and oversight, international trade issues, and policy concerns. The report focuses on five technologies in particular: embryo transfer, in vitro fertilization, sexing embryos, transgenics, and cloning. This report was updated at least once a year until 2010.

Fernandez-Cornejo, Jorge, et al. 2014. "Genetically Engineered Crops in the United States." Report ERR-162. U.S. Department of Agriculture. Economic Research Service.

> This report traces the adoption and use of genetically modified crops in the United States from 1996 through 2014 with a consideration of possible effects on the environment caused by such crops.

Friends of the Earth, International. 2009. "Who Benefits from GM Crops? Feeding the Biotech Giants, Not the World's Poor." Amsterdam: Friend of the Earth.

> This report presents a very detailed, well-documented case that the greatest beneficiaries of the production of GM foods are not consumers, but food production companies. The report presents statistical data on a host of GM-related issues, including GM products now in production and use; percentage of land devoted to GM crop production; amount and cost of pesticides used in GM crop production; average cost of various GM foods in the United States and other nations; and crop yields for GM and conventional foods.

Holtzman, Neil A., and Michael S. Watson. 1998. "Promoting Safe and Effective Genetic Testing in the United States: Final Report of the Task Force on Genetic Testing." Baltimore: Johns Hopkins University Press.

> The Task Force on Genetic Testing was created in April 1995 by the National Institutes of Health-Department of Energy Working Group on Ethical, Legal, and Social Implications of Human Genome Research. Its charge was to review genetic testing in the United States and to make recommendations that would ensure the development

of safe and effective genetic tests. The six chapters in the report deal with a general introduction to the technology and issues raised by its use, ensuring the safety and efficacy of new genetic testing procedures, ensuring the proficiency of testing laboratories, improving the understanding of the technology by health care providers, special problems involved in testing for rare genetic disorders, and summary and recommendations.

"Human Gene Transfer Research." 2011. National Reference Center for Bioethics Literature. Joseph and Rose Kennedy Institute of Ethics. Georgetown University. https://repository.library.georgetown.edu/bitstream/handle/10822/551524/scope_note24.pdf?sequence=1&isAllowed=y. Accessed on April 22, 2016.
   This report was first published in March 1994 and was updated regularly through 2011. It provided a comprehensive overview of the field of human gene therapy in a format easily comprehendible to the average reader. It included a general introduction to the topic, a history of gene therapy in the United States, a discussion of techniques, arguments in favor of and opposed to gene therapy, and a list of organizations and references on the topic of gene therapy.

James, Clive. 2015. "Global Status of Commercialized Biotech/GM Crops: 2014." ISAAA Brief No. 51. Ithaca, NY: International Service for the Acquisition of Agri-Biotech Applications. http://www.isaaa.org/resources/publications/briefs/51/default.asp. Accessed on April 22, 2016.
   This report is one of many such reports produced by the ISAAA on the status of genetically engineered crops worldwide.

National Commission on the Future of DNA Evidence. 2000. "The Future of Forensic DNA Testing." Washington, DC: U.S.

Department of Justice. Office of Justice Programs. National Institute of Justice.

This report was written by a committee appointed by then-attorney general Janet Reno to consider future applications of DNA typing in the United States. The report consists of two parts, the first of which is a general introduction to the subject and a consideration of possible future applications of DNA typing written for the general public. The second part of the report consists of a series of appendices that deal with technical issues involved in the use of DNA typing in forensic science.

National Commission on the Future of DNA Evidence. 1999. "What Every Law Enforcement Officer Should Know about DNA Evidence: A Pocket Guide." Washington, DC: U.S. Department of Justice. Office of Justice Programs. National Institute of Justice.

This short pamphlet (described as a "report" by the commission) provides basic information about DNA typing, circumstances in which it can and should be used in law enforcement activities, and special cautions to observe in its use.

National Institute of Justice. 2003. "Report to the Attorney General on Delays in Forensic DNA Analysis." Washington, DC: U.S. Department of Justice. Office of Justice Programs.

One of the most difficult problems facing the use of DNA evidence in forensic science is the length of time required to obtain analyses of evidence collected at a crime scene. Wait times of many months are not unusual for prosecutors and other law enforcement officers hoping to use such data in crime investigations and prosecutions. This report analyzes the current status of delays in DNA testing and interpretation and changes needed in the future to make this technology more useful to the criminal justice system.

Nuffield Council on Bioethics. "The Forensic Use of Bioinformation: Ethical Issues." London: Nuffield Council on Bioethics, September 2007.

This report addresses a number of current issues related to the use of fingerprints, DNA evidence, and other types of bioinformation by law enforcement agencies in the United Kingdom. It deals with issues such as the circumstances under which police officers should be permitted to take digital and DNA fingerprints, how long those prints and the profiles made from them should be retained, how bioinformation should and should not be used in criminal trials, and what the proper role of a national DNA database is.

Office of Science and Technology Policy. 1986. "Coordinated Framework for Regulation of Biotechnology." Federal Register. 51: 23302.

This document is of primary importance because it sets out U.S. policy on the regulation of all forms of biotechnology, explaining which agencies are responsible for which fields of research and laying out the policies and regulations of each agency with regulatory responsibility. Available online at https://www.aphis.usda.gov/brs/fedregister/coordinated_framework.pdf. Accessed on April 17, 2016.

Organisation for Economic Co-operation and Development. 2007. "Genetic Testing: A Survey of Quality Assurance and Proficiency Standards." Paris: Organisation for Economic Co-operation and Development.

The report summarizes the results of a survey conducted with 800 genetic testing laboratories in 18 countries in the European Union. It provides information on laboratory policies and practices and the proficiency of workers in those laboratories; policies regarding samples and genetic data handling; and the transborder flow of specimens.

Pew Initiative on Food and Biotechnology. 2003. "Pharming the Field: A Look at the Benefits and Risks of Bioengineering Plants to Produce Pharmaceuticals." Washington, DC: The Pew Charitable Trusts.

This document reports on the results of a workshop sponsored by the Pew Initiative on Food and Biotechnology, the U.S. Food and Drug Administration and the Cooperative State Research, Education and Extension Service of the U.S. Department of Agriculture held in Washington in July 2002. The report summarizes potential risks and benefits associated with the technology, along with regulatory and legal issues that arise as a result of its use in the production of pharmaceuticals.

President's Council on Bioethics. 2008. "The Changing Moral Focus of Newborn Screening." Washington, DC: [The President's Council on Bioethics].

Each year, about four million newborn children in the United States are screened for one or more diseases. Technology has made possible screening for a much wider range of diseases, some of which are curable and some are not. This change in technology prompted the President's Council on Bioethics to consider public policy decisions that should be implemented to help parents and health providers to decide which tests under which conditions should be made available to the parents of newborn children.

Roman, John K., et al. 2008. "The DNA Field Experiment: Cost-Effectiveness Analysis of the Use of DNA in the Investigation of High-Volume Crimes." Washington, DC: Urban Institute. Justice Policy Center.

This report deals with the issue of DNA testing in so-called high-volume crimes, such as burglary, for which DNA typing has historically not been applied. The report summarizes evidence obtained from research conducted in five urban areas: Phoenix, Arizona; Orange County, California; Denver, Colorado; Los Angeles, California; and Topeka,

Kansas. It finds substantial reasons for making use of DNA typing in "minor" crimes as well as in murder, aggravated assault, rape, and other so-called "major" crimes.

Sarata, Amanda K. 2015. "Genetic Testing: Background and Policy Issues." Washington, DC: Library of Congress. Congressional Research Service.

This report is prepared for the U.S. Congress. It reviews the current status of genetic testing in the United States and outlines policy issues that have arisen in the areas of the type of information that genetic tests can provide, the validity and evaluation of genetic tests, and the use of genetic test results.

Internet

Adenle, Ademola A. 2011. "Response to Issues on GM Agriculture in Africa: Are Transgenic Crops Safe? *BioMedCentral*. http://bmcresnotes.biomedcentral.com/articles/10.1186/1756-0500-4-388. Accessed on April 18, 2016.

The author addresses concerns about the introduction of genetically modified crops into Africa because of potential safety concerns. He argues that food shortages on the continent are so severe that a generous attitude toward the contribution that GM foods can make to the crisis should be observed.

Baer, Drake. 2015. "This Korean Lab Has Nearly Perfected Dog Cloning, and That's Just the Start." Tech Insider. http://www.techinsider.io/how-woosuk-hwangs-sooam-biotech-mastered-cloning-2015-8. Accessed on April 23, 2016.

The author provides a thorough and interesting introduction to the science of pet cloning, with its advantages and disadvantages.

Brand, Stewart. 2013. "The Dawn of De-Extinction: Are You Ready?" TED. https://www.ted.com/talks/stewart_brand_the_dawn_of_de_extinction_are_you_ready?language=en. Accessed on April 23, 2016.

This TED talk is based on the presumption that the ability to restore extinct animals is closer than most people realize, and questions as to whether or not, or with which species, that process should be allowed to take place.

Budowle, Bruce, Arthur J. Eisenberg, and Angela van Daal. 2009. "Validity of Low Copy Number Typing and Applications to Forensic Science." *Croatian Medical Journal.* http://www.denverda.org/DNA_Documents/CMJ%20Budowle.pdf. Accessed on April 22, 2016.

This article presents an excellent overview of the new field of low copy number DNA analysis with a discussion of its applications in forensic science and a number of issues relating to its effective use in the field.

Caplan, Arthur. 2015. "GMO Foods Should Be Labeled, But Not for Safety: Bioethicist." *NBC News.* http://www.nbcnews.com/health/health-news/why-gmo-foods-should-be-labeled-n423451. Accessed on April 22, 2016.

Caplan argues that safety is not an issue for GM foods, so that should not be the reason to require that they be labeled. Labeling, he says, is needed simply because people have a right to know the components of the food they eat.

Chassy, Bruce, and Jon Entine. 2015. "Although Some GMO Sympathizers Embrace Mandatory Labeling, It's a Disaster in Waiting." *Huffpost Science.* http://www.huffingtonpost.com/jon-entine/although-some-gmo-sympath_b_8864038.html. Accessed on April 22, 2016.

This article is the first in a group of three articles explaining the authors' opposition to the labeling of GM foods in considerable detail. For the other two articles, see "The Real Cost of Mandatory GMO Labeling" (http://www.huffingtonpost.com/jon-entine/the-real-cost-of-mandato r_b_8865742.html) and "Why We Oppose GMO Labeling: Science and the Law" (http://www.huffingtonpost.com/jon-entine/gmo-labeling-science-and-_b_8871680.html).

Chavil, Bobby, Annette Summers, and Mary Napier. 2013. "DNA Fingerprinting Comes of Age." *Forensic Magazine.* http://www.forensicmag.com/articles/2013/08/dna-finger printing-comes-age. Accessed on April 22, 2016.

> This article presents a very good general overview of the current status of DNA fingerprinting today, with a discussion of the technology used to analyze DNA, the DNA typing system used in the United States, and future prospects for the technology.

"Combined DNA Access System (CODIS)." 2016. Federal Bureau of Investigation. https://www.fbi.gov/about-us/lab/ biometric-analysis/codis. Accessed on April 17, 2016.

> This Web site provides an introduction to the FBI CODIS system, with additional detailed information on a variety of related topics, such as an explanation as to how the system works, statistical information, and description of some applications of DNA forensic analysis.

Cottrell, Sariah, Jamie L. Jensen, and Steven L. Peck. 2014. "Resuscitation and Resurrection: The Ethics of Cloning Chee-tahs, Mammoths, and Neanderthals." *Life Sciences, Society and Policy.* 10: 3. doi: 10.1186/2195-7819-10-3. http://lssp journal.springeropen.com/articles/10.1186/2195-7819-10-3. Accessed on April 23, 2016.

> The authors provide a detailed discussion of the technol-ogy and ethical issues involved in the cloning of endan-gered and extinct species.

Das, Ranajit, et al. 2016. "Localizing Ashkenazic Jews to Pri-meval Villages in the Ancient Iranian Lands of Ashkenaz." *Genome Biology and Evolution.* http://gbe.oxfordjournals.org/ content/8/4/1132. Accessed on April 20, 2016.

> Researchers explain how they used DNA of 393 living Ashkenazic, Iranian, and mountain Jews to determine the place of origin of the Yiddish language.

"De-Extinction." 2016. *National Geographic.* http://www.nation algeographic.com/deextinction/. Accessed on April 23, 2016.

This Web site provides a number of different articles on various aspects of the science, technology, ethics, and morality of restoring extinct species to life.

Dehghan, Saeed Kamali. 2015. "Scientists in Iran Clone Endangered Mouflon—Born to Domestic Sheep." *The Guardian.* https://www.theguardian.com/science/2015/aug/05/iran-scientists-clone-endangered-mouflon-domestic-sheep. Accessed on April 23, 20167.

This news article reports on the cloning of an endangered animal, the mouflon, and describes the process by which the research was done.

"DNA Evidence Basics." 2012. National Institute of Justice. http://nij.gov/topics/forensics/evidence/dna/basics/pages/welcome.aspx. Accessed on April 17, 2016.

This Web site provides a superb overview of virtually all aspects of DNA forensic testing, including the basics of collecting and transporting DNA evidence, the analysis of DNA evidence, possible results from a DNA test, and types of samples that are suitable for DNA testing.

"DNA Work Saves Fish from Extinction." 2016. Phys.org. http://phys.org/news/2016-03-dna-fish-extinction.html. Accessed on April 20, 2016.

This article explains how the determination of the complete genome of the Yarra pygmy perch (*Nannoperca obscura*) and southern pygmy perch (*N. australis*) made it possible to develop programs most likely to increase the survival of these endangered fish in Australia.

"Frequently Asked Questions about Gene Testing." 2015. National Human Genome Research Institute. https://www.genome.gov/19516567/faq-about-genetic-testing/. Accessed on April 18, 2016.

This Web page provides a readable general introduction to the subject of gene testing, including a definition of the procedure, an explanation as to how it is carried out, some pros and cons of testing, information on regulation, insurance coverage for the procedure, and links to other print and electronic resources.

"GE Food Labeling." 2016. Center for Food Safety. http://www.centerforfoodsafety.org/issues/976/ge-food-labeling/about-ge-labeling. Accessed on April 22, 2016.

This Web site provides a host of information in favor of the labeling of GM foods, including reasons for supporting the practice, public opinion polls on the subject, existing state legislation, proposed federal legislation, legal actions, and related topics.

"Genetic Testing." 2016. Mayo Clinic. http://www.mayoclinic .org/tests-procedures/genetic-testing/basics/definition/prc-20014802. Accessed on April 18, 2016.

This highly respected electronic resource provides an overview of genetic testing for the layperson, focusing on a discussion as to the conditions under which the procedure is used, how a patient prepares for a test, what results might be expected, and how those results are to be assessed and interpreted.

"Genetic Testing." 2016. U.S. National Library of Medicine. https://ghr.nlm.nih.gov/primer/testing.pdf. Accessed on April 18, 2016.

This publication provides an excellent overview of the topic of genetic testing, with sections on types of testing, testing methods, the meaning of informed consent, interpreting genetic test results, benefits of genetic testing, insurance coverage for the procedure, and other related topics.

Gyngell, Chris, Tom Douglas, and Julian Savulescu. 2015. Bioethics Blog Tracker. https://bioethics.georgetown.edu/2015/12/engineering-a-consensus-edit-embryos-for-research-not-repro duction/. Accessed on April 21, 2016.

The authors note the availability of new gene-editing technologies (such as CRISPR-Cas9) that once more raise issues about the cloning of humans and other animals. They review the moral and social arguments for and against such procedures and conclude that the existing policy of allowing therapeutic, but banning reproducing, cloning remains the best policy for the near future.

Harris, Dennis. 2013. "New DNA Advances." *Evidence Technology* magazine. http://www.evidencemagazine.com/index.php?option=com_content&task=view&id=1114. Accessed on April 22, 2016.

The author focuses on three important advances in the use of DNA fingerprinting in forensic science: expanded and improved use of databases, extension of databases by adding more arrestee profiles, and greater use of new rapid DNA analysis technology.

Hawana, Joanne S. 2015. "FDA Finalizes Genetically Engineered Food Labeling Guidance & Approves 'AquAdvantage Salmon'." *National Law Review*. http://www.natlawreview.com/article/fda-finalizes-genetically-engineered-food-labeling-guidance-approves-aquadvantage. Accessed on April 21, 2016.

This article provides a review of the process by which AquAdvantage salmon was approved and some of the social and legal problems that have been and are associated with this decision.

Heintzman, Peter D. 2013. "Patterns in Palaeontology: An Introduction to Ancient DNA." *Paleontology* [online]. 3(10): 1–10. http://www.palaeontologyonline.com/articles/2013/patterns-in-palaeontology-an-introduction-to-ancient-dna/. Accessed on April 22, 2016.

The author explains how DNA is extracted from fossil remains, the information that can be obtained from the DNA, and future prospects for the use of DNA in future palaeontological research.

Jabr, Ferris. 2013. "Will Cloning Ever Save Endangered Animals?" *Scientific American.* http://www.scientificamerican.com/ article/cloning-endangered-animals/. Accessed on April 23, 2016.
  The author notes that the technology for saving endangered animals with cloning technology is not yet available, but it is likely to be so in the near future. He asks what the pros and cons of proceeding with such experiments are.

Kayser, Manfred, and Peter de Knijff. 2011. "Improving Human Forensics through Advances in Genetics, Genomics and Molecular Biology." *Nature Reviews Genetics.* http://www.nist.gov/ mml/bmd/genetics/upload/nrg2952.pdf. Accessed on April 22, 2016.
  The authors discuss advances in the field of genetics and molecular biology that have had implications on forensic sciences, such as the development of new types of genetic markers, new methods of identifying suspects previously unknown to investigators, and new methods for establishing links between DNA donors and crime suspects.

Knapp, Michael, Carles Lalueza-Fox, and Michael Hofreiter. 2015. "Re-inventing Ancient Human DNA. 2015. *Investigative Genetics.* doi: 10.1186/s13323-015-0020-4. http:// link.springer.com/article/10.1186%2Fs13323-015-0020-4. Accessed on April 20, 2016.
  The authors note that using DNA in the study of ancient humans has long been an issue of considerable debate. They then point out that improvements of the fields have been such as to make the technology "a central component of modern anthropological research."

Knoepfler, Paul. 2015. "Are Babies from Same-Sex Couples Really Possible?" Knoepfler Lab Stem Cell Blog. https://www.ipscell.com/ 2015/03/babysamesexcouple/#. Accessed on April 21, 2016.

Knoepfler reviews a number of recent articles about the possibility of same-sex couples having babies (links are provided to the articles) and discusses how realistic such possibilities are.

"Kuwait Set to Enforce DNA Testing Law on All—Officials Reassure Tests Won't Be Used to Determine Genealogy." *Kuwait Times.* http://news.kuwaittimes.net/website/kuwait-to-enforce-dna-testing-law-on-citizens-expats-visitors-tests-wont-be-used-to-determine-genealogy-affect-freedoms/. Accessed on April 20, 2016.

On July 2, 2015, the Kuwaiti legislature passed a new law requiring everyone in Kuwait—both residents and visitors—to have their DNA tested for possible use in fighting crime and terrorism. The government assured people that DNA information would not be made available for any other purposes, although some critics have questioned whether that promise can be kept. Kuwait was the first country in the world to adopt such a law.

"Making a Transgenic Animal." 2010. Biotechnology Learning Hub.

This animated program describes the steps involved in the production of a transgenic cow, with detailed additional explanations for each step in the process.

McDonnell, W. Michael, and Frederick K. Askari. 2016. "The Emerging Role of DNA Vaccines." *Medscape.* http://www.medscape.com/viewarticle/715527_1. Accessed on April 23, 2016.

This article provides a very complete description of DNA vaccines and their potential to replace or compliment traditional vaccine in future health programs.

Mikadze, Kirsten. 2016. "The GM Salmon Debate: Allow, Label, or Outlaw?" Siskinds. http://envirolaw.com/the-gm-salmon-debate-allow-label-or-outlaw/. Accessed on April 21, 2016.

An environmental law attorney explores the legal issues related to the development of the first genetically modified animal, the AquAdvantage salmon.

Morella, Cecil. 2016. "DNA Rice Breakthrough Raises 'Green Revolution' Hopes." Phys.org.http://phys.org/news/2016-02-dna-rice-breakthrough-green-revolution.html. Accessed on April 20, 2016.
This article describes progress in research on the DNA of rice, and the potential that research has for the improvement of rice crops in the future.

Murray, James D., and Jenny Graves. 2016. "Opposition to Genetically Modified Animals Could Leave Millions Hungry." The Conversation. http://theconversation.com/opposition-to-genetically-modified-animals-could-leave-millions-hungry-53740. Accessed on April 21, 2016.
The authors argue that unreasonable objections to the development and use of genetically modified foods pose a threat to hundreds of millions of people worldwide who regularly do not get enough food to eat every day.

"Recombinant DNA and Gene Cloning." 2014. http://www.biology-pages.info/R/RecombinantDNA.html. Accessed on April 18, 2016.
This Web site provides an excellent, succinct, well-presented general introduction to the process by which recombinant DNA procedures are carried out. The presentation is technical, but at a level that most individuals should be able to understand.

Regalado, Antonio. 2016. "DNA App Store." *MIT Technology Review.* https://www.technologyreview.com/s/600769/10-breakthrough-technologies-2016-dna-app-store/. Accessed on March 22, 2016.
This annual review discusses the 10 most important breakthroughs in DNA technology each year, with an introduction to each technology.

Regalado, Antonio. 2015. "First Gene-Edited Dogs Reported in China." *MIT Technology Review.* https://www.technolo gyreview.com/s/542616/first-gene-edited-dogs-reported-in-china/. Accessed on April 23, 2016.

The author reports on the first use of new gene-editing technology (CRISPR-Cas9) for the production of genetically modified dogs at a laboratory in China, with a discussion of the possible significance of this research for similar types of research with other animals.

Regalado, Antonio. 2015. "Surgeons Smash Records with Pig-to-Primate Organ Transplants." *MIT Technology Review.* https://www.technologyreview.com/s/540076/surgeons-smash-records-with-pig-to-primate-organ-transplants/. Accessed on April 23, 2016.

This article describes an important step forward in the field of xenotransplantation with the 945-day survival of a baboon carrying a genetically engineered pig heart, encouraging optimism about the greater likelihood of such procedures with humans.

Riley, Donald E. 2005. "DNA Testing: An Introduction For Non-Scientists; An Illustrated Explanation." http://www.scien tific.org/tutorials/articles/riley/riley.html. Accessed on April 18, 2016.

This Web site provides a very complete and detailed explanation of the technology used in genetic testing. It is written for the general public, although some background in chemistry would be useful for a good understanding of the subject matter. Probably an essential resource for anyone who wishes to understand the technical basis of genetic testing.

Riordan, Sean M., et al. 2015. "Application of CRISPR/Cas9 for Biomedical Discoveries. Cell & Bioscience. doi: 10.1186/ s13578-015-0027-9. http://link.springer.com/article/10.1186/ s13578-015-0027-9. Accessed on April 20, 2016.

The authors provide a review of CRISPR-Cas9 system for genome editing, explain some of its possible applications in the field of biomedicine, and suggest some possible future directions for this line of research.

Roewer, Lutz. 2013. "DNA Fingerprinting in Forensics: Past, Present, Future." *Investigative Genetics.* http://download.springer .com/static/pdf/912/art%253A10.1186%252F2041-2223-4-22 .pdf?originUrl=http%3A%2F%2Finvestigativegenetics.biomed central.com%2Farticle%2F10.1186%2F2041-2223-4-22&token 2=exp=1461360386~acl=%2Fstatic%2Fpdf%2F912%2Fart% 25253A10.1186%25252F2041-2223-4-22.pdf*~hmac=973ace 8d4924933e4ab7a533cdf099ca6003749349b8076113f1bc59b 93635f3. Accessed on April 22, 2016.

This article offers a very clear, accurate, and nicely illustrated recital of the history of DNA fingerprinting and its contributions to forensic science over the past 30 years.

Romeika, Jennifer M., and Fei Yan. 2013. "Recent Advances in Forensic DNA Analysis." *Forensic Research.* http://www .omicsonline.org/recent-advances-in-forensic-dna-analysis-2157-7145.S12-001.pdf. Accessed on April 22, 2016.

The DNA technology used in forensic science constantly evolves, with new discoveries and updates of existing procedures. This review article summarizes and discusses some of the most recent of those advances.

Samdani, Tushar. 2014. "Xenotransplantation." *Medscape.* http:// emedicine.medscape.com/article/432418-overview#a1. Accessed on April 23, 2016.

This well-written article provides a comprehensive overview of xenotransplantation, with sections on the biological and immunological problems and other medical and ethical issues associated with the practice.

Seddon, Steve. 2015. "Recombinant DNA." YouTube. https:// www.youtube.com/watch?v=8Dd7M9PGhgQ. Accessed on April 18, 2016.

This video is one of many available on the Internet that provides a nice animated introduction to the process by which recombinant DNA changes are carried out.

Seruggia, Davide, and Lluís Montoliu. 2016. "The CRISPR Page at CNB." http://wwwuser.cnb.csic.es/~montoliu/CRISPR/. Accessed on April 4, 2016.
This Web page is a treasure chest of information on all aspects of the CRISPR system, including an explanation of the system, its history, and essential papers on the method for gene editing.

Signorile, Michelangelo, and Alondra Nelson. 2016. "How Do Advancements in DNA Testing Technology Help the Reparations Movement?" NewBlackMan (in Exile). http://www .newblackmaninexile.net/2016/03/how-do-advancements-in-dna-testing.html. Accessed on April 20, 2016.
This radio interview focuses on the way in which research in DNA testing has provided a new and very useful tool for those who are interested in providing reparations to African Americans today for the institution of slavery perpetrated on their ancestors in the United States.

Stein, Rob. 2015. "Cloning Your Dog, for a Mere $100,000." NPR. http://www.npr.org/sections/health-shots/2015/09/30/428927516/cloning-your-dog-for-a-mere-100-000. Accessed on April 23, 2016.
This article provides a good overview of the potential for having pets cloned by modern technological methods.

Swetlitz, Ike. 2016. "FDA Urged to Approve 'Three-Parent Embryos,' a New Frontier in Reproduction." STAT. https://www .statnews.com/2016/02/03/three-parent-embryos-reproduction/. Accessed on April 22, 2016.
This article provides an explanation of mitochondrial replacement technology, a method for correcting genetic

errors in women who are unable to bring pregnancies to a successful conclusion.

"Transgenic Crops: An Introduction and Resource Guide." 2004. Colorado State University. http://cls.casa.colostate.edu/transgeniccrops/current.html. Accessed on April 18, 2016.
Although somewhat dated in some regards, this Web site provides a vast amount of very useful information about transgenic plants written in a form easily available to most readers.

"Transgenic Plants." 2016. Nature.com. http://www.nature.com/subjects/transgenic-plants. Accessed on April 18, 2016.
This Web page lists recent articles that have appeared in the journal *Nature* on the topic of transgenic plants. The articles are both technical in nature and aimed at the general reader.

United Mitochondrial Disease Foundation. 2016. http://www.umdf.org/site/c.8qKOJ0MvF7LUG/b.9166823/k.2E25/Mitochondrial_Replacement_Therapy.htm. Accessed on April 22, 2016.
This web site provides up-to-date information on mitochondrial replacement therapy, a method for treating a condition in women that may prevent their bringing a pregnancy to successful birth.

"What Are the Pros and Cons of Transgenic Crops?" 2016. The Maize Full Length cDNA Project. http://www.maizecdna.org/outreach/tpe.html. Accessed on April 18, 2016.
This Web site provides an excellent and detailed review of the arguments generally put forward both for and against the development and use of transgenic crops.

"What Is DNA Barcoding." 2016. Barcode of Life. http://www.barcodeoflife.org/content/about/what-dna-barcoding. Accessed on April 20, 2016.

This article describes a new application of DNA technology, the use of sections of an organism's DNA to assign a taxonomic status to that organism. The article describes the scientific theory on which the application is based and some of the ways it has been and can be applied in making taxonomic decisions.

Word, Charlotte. 2010. "What Is LCN? Definitions and Challenges." Promega. http://www.promega.com/resources/profiles-in-dna/2010/what-is-lcn-definitions-and-challenges/. Accessed on March 23, 2016.

Low copy number (or low count number) DNA analysis is a highly precise method for identifying the DNA of very small amounts of genetic material. It has been widely used for more than a decade, but it has been subjected to some criticism for the accuracy of its results, as discussed in this article.

Zimmer, Carl. 2016. "DNA under the Scope, and a Forensic Tool under a Cloud." *New York Times*. http://www.nytimes.com/2016/02/27/science/dna-under-the-scope-and-a-forensic-tool-under-a-cloud.html?_r=0. Accessed on March 22, 2016.

The author outlines some of the problems involved with the development and use of ever more precise methods of DNA forensic analysis.

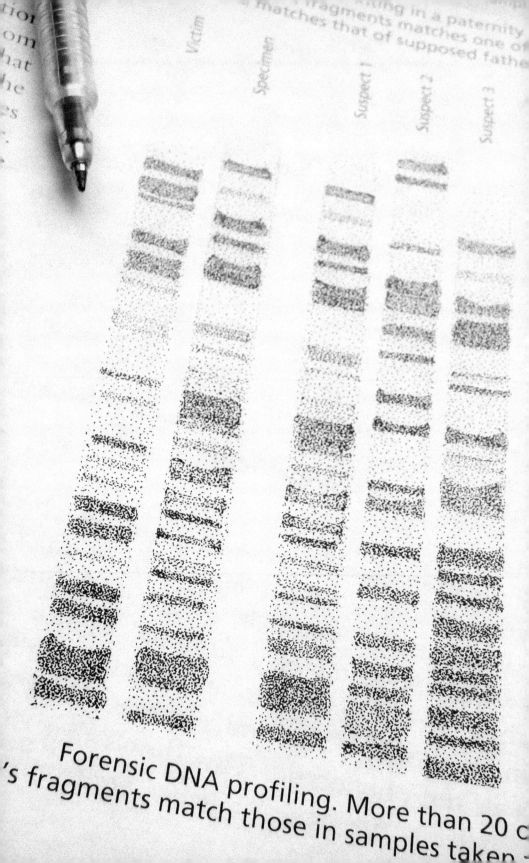

Forensic DNA profiling. More than 20 c
's fragments match those in samples taken

This chapter provides a time line of events in the history of recombinant DNA research and its applications. It includes important scientific and technological developments, as well as social, political, and ethical issues related to DNA technology. The dates in this chapter are often approximate since, for example, a researcher may begin his or her studies in one year, continue them for many years, and report results at some later date or dates. Or, a law may be passed in one year, but not take effect until sometime later.

Premodern Times   Humans discover any number of basic principles in biotechnology for which they have no scientific explanation. These principles and processes arise as a result of trial-and-error efforts in the fields of agriculture, dairying, beer and wine production, fermentation, food preservation, and human growth patterns.

1865   Gregor Mendel, an Austrian monk, presents his research on growth patterns in peas to the Natural Science Society of Brünn. He hypothesizes that characteristic traits are transmitted from one generation to the next by means of certain "factors" contained within plants. Mendel's so-called "factors" are now known as genes.

These DNA "fingerprints" are similar to bar codes on commercial products that provide a distinct identification for a person's identity. (Joanne Zh/Dreamstime.com)

1868   Charles Darwin hypothesizes that the unit of inheritance in plants and animals is a unit that he calls the gemmule. He is, at the time, unaware of Mendel's research or that his "gemmule" is similar both to Mendel's "factor" and to the modern-day gene.

1869   Swiss physician and biologist Johannes Friedrich Miescher discovers a previously unknown phosphorus-rich compound which he calls *nuclein*. The compound is later found to be a nucleic acid, although its biological function is not understood for many years.

1879   German physician and researcher Albrecht Kossel begins his research on the chemical composition of cellular components. He discovers that Miescher's nuclein is actually a class of biochemical compounds later given the name of by German pathologist Richard Altman in 1889.

1883   English polyglot Francis Galton coins the term *eugenics* to improve the human race by selective breeding.

1884   Mendel dies without his accomplishments having been recognized in his lifetime.

1900   Three scientists, Dutch botanist Hugo de Vries, Austrian botanist Erich von Tschermak, and German botanist Carl Correns, independently rediscover Mendel's research in preparing for their own genetic studies.

1902   American geneticist and physician Walter Stanborough Sutton proposes the chromosomal theory of inheritance. He says that Mendel's "factors" are units located on chromosomes, which can explain his theory of independent assortment of genetic characteristics.

1903   Danish botanist and geneticist Wilhelm Johannsen suggests the terms *gene, genotype*, and *phenotype* to describe the unit of inheritance and the genetic and physical features of an organism produced by the transmission of genes.

1906   British geneticist William Bateson suggests the word *genetics* to describe the science of inheritance.

1908   British physician and medical researcher Archibald Edward Garrod proposes the one gene/one enzyme theory, which says that each gene is responsible for the production of one specific protein (enzyme). Garrod's idea is largely ignored until it is rediscovered and proved by the research of American geneticists George Beadle and Edward Tatum in 1941.

1909   Russian American biochemist Phoebus Levene discovers the presence of the sugar ribose in nucleic acids. Nucleic acids containing this compound later become known as *ribonucleic acids*, or *RNA*.

1911   American geneticist and embryologist Thomas Hunt Morgan establishes the role of genes and chromosomes in the transmission of inheritable characteristics in a series of ingenious experiments with fruit flies.

1913   American geneticist Alfred Sturtevant creates the first genetic map of a chromosome, one of the fruit fly's eight chromosomes.

1917   Hungarian engineer Karl Ereky coins the term *biotechnology* and uses the term to refer to any product made from raw materials with the aid of living organisms.

1922   American geneticist Hermann J. Muller publishes a remarkably prescient paper, "Variation Due to Change in the Individual Gene," in which he summarizes current knowledge about the chemical and physiological characteristics of the gene and its role in the inheritance of physical characteristics. At this point, however, scientists do not yet know what a gene is.

1929   Levene discovers a previously unknown component of nucleic acids, the sugar deoxyribose, in place of the sugar ribose already known to be a component of nucleic acids. He calls the new kind of nucleic acid *deoxyribose nucleic acid*, or *DNA*.

1934   British physicist John Desmond Bernal finds that X-ray crystallography can be a powerful tool in determining the molecular structure of proteins. The method later becomes an essential tool for finding the molecular structure of DNA.

1935 Russian biochemist Andrei Nikolaevitch Belozersky isolates DNA in the pure state for the first time.

1938 American engineer, mathematician, and science administrator Warren Weaver suggests the name *molecular biology* for the new field of science in which biological phenomena are studied and interpreted in chemical and physical (primarily molecular) terms.

1940 Belgian biochemist Jean Brachet hypothesizes that DNA and RNA in cells have two very different functions, and that the latter is actually involved in the production of protein molecules. Brachet's hypothesis presages the fundamental dogma of molecular biology, namely that DNA makes RNA, which makes proteins.

1941 American geneticists George Beadle and Edward Tatum rediscover and prove A. E. Garrod's one gene/one enzyme theory (1908).

Danish microbiologist A. Justin (also given as Jost and Joost) is credited with first having used the term *genetic engineering* to describe the transfer of one piece of genetic material from one organism to a second organism.

1944 Medical researchers Oswald Avery, Colin MacLeod, and Maclyn McCarty hypothesize that DNA is responsible for transformations observed in succeeding generations of bacteria. They argue that DNA may play the role previously assigned to genes, in more complex organisms. The hypothesis is startling and largely rejected, because researchers are largely convinced that nucleic acids are too small and simple to carry genetic information (and that proteins are the more likely candidates for this role).

1946 German American biophysicist Max Delbrück and American bacteriologist and geneticist Alfred Day Hershey independently find that the genetic material from two different viruses can combine with each other in such a way as to produce a third and new type of virus. Their experiments are an early example of the recombination of DNA that lies at the basis of much of DNA technology today.

1950   Austrian American biochemist Erwin Chargaff discovers that the quantity of nitrogen bases adenine and thymine is equal to that of the bases cytosine and guanine. Although there is no obvious significance to this discovery at the time, it provides a key piece of information needed by Watson and Crick in their elucidation of the structure of the DNA molecule (1953).

British biophysicist Rosalind Franklin and New Zealand–born British molecular biologist Maurice Wilkins begin their X-ray crystallographic studies of DNA molecules and conclude that they exist as double-stranded helices, information that became a key to the solution of DNA structure by Watson and Crick in 1953.

1952   American molecular biologists Joshua Lederberg and Norton Zinder discover the process of bacterial transduction, in which a virus removes DNA from one bacterial cell and transports it to a second bacterium. The process, in principle, is one used by modern-day researchers to alter the genetic constitution of a cell, and it is still called by that name.

Hershey and American geneticist Martha C. Chase (Epstein) carry out the famous "blender experiment," in which they conclusively demonstrate that DNA (and not protein) is the genetic material.

American biologists Robert Briggs and Thomas King clone the first vertebrate by transplanting nuclei harvested from leopard frogs into eggs from which nuclei had been removed.

1953   American biologist James Watson and English chemist Francis Crick elucidate the structure of the DNA molecule, showing that it occurs as a double-stranded helical chain of sugar and phosphate groups linked by four nitrogen bases.

1955   American physicist, molecular biologist, and behavioral geneticist Seymour Benzer develops methods for studying the detailed molecular characteristics of mutations and finds, among other things, that changes can take place at specific points within a DNA molecule, sometimes at the level of a single nucleotide.

1957   American geneticist and molecular biologist Matthew Meselson and American molecular biologist Franklin Stahl demonstrate the process by which DNA molecules self-replicate, a process described as *semiconservative*, because the two strands of the DNA molecule act as templates for the construction of two new identical DNA molecules without being destroyed.

1958   Francis Crick enunciates the Central Dogma principle of molecular biology, essentially that "once information has gotten into a protein it can't get out again." The Central Dogma further says that the most common type of information transfer in a cell is DNA → RNA → Protein, but under special circumstances, may include RNA → DNA → Protein.

1959   French geneticist Jérôme Lejeune demonstrates that Down syndrome results from the presence of an extra copy of all or part of chromosome number 21, the first instance in which a genetic basis for a disease had been identified.

1961   American microbiologist Robert Guthrie develops a method for screening the blood of a newborn child for the genetic disease phenylketonuria.

1966   Indian American molecular biologist Har Gobind Khorana and American biochemists Marshall Nirenberg and Robert Holley elucidate the genetic code, the set of three nitrogen bases that code for the amino acids that make up proteins.

1968   Swiss microbiologist Werner Arber discovers restriction enzymes, compounds with the ability to recognize and cut DNA molecules at specific points.

1970   American molecular biologist David Baltimore and American geneticist Howard Temin independently discover the enzyme reverse transcriptase, which facilitates the transcription of RNA material into a new DNA molecule, a discovery with a host of applications in molecular biology, one of which is a method for producing clones.

American microbiologist Hamilton O. Smith discovers the first site-specific restriction enzyme, a compound that recognizes a specific base sequence in a DNA molecule and then cuts the molecule at some point within that sequence.

1972   American biochemist Paul Berg develops a method for inserting a segment of DNA from a foreign molecule into a host DNA molecule, producing the first recombinant DNA molecule. The method is sometimes referred to as *gene splicing*. Berg and some colleagues soon write to the National Institutes of Health pointing out possible safety, social, and ethical issues related to this research, an act that eventually leads to the Asilomar Conference of 1975.

1973   American biochemists Stanley N. Cohen and Herbert Boyer develop a method for inserting foreign DNA segments into a plasmid and then incorporating the altered plasmid molecule into the DNA of a bacterium. The procedure results in the formation of an engineered organism (the bacterium) that can then make endless copies of itself (cloning), while displaying characteristics different from those of the original bacterium. The experiment marks the beginning of the modern field of recombinant DNA procedures.

1974   The National Institutes of Health (NIH) establishes the Recombinant DNA Advisory Committee to provide federal oversight for all recombinant DNA research. The committee continues to function today as a forum for scientific, ethical, and legal issues raised by recombinant DNA technology and its research and clinical applications.

German American biochemist Rudolf Jaenisch produces the first transgenic animal when he injects a cancer gene into mouse embryos. At maturity, the mice born from this experiment carry the cancer gene on their own DNA. The gene is transmitted, along with the rest of the mouse genome, to succeeding generations.

1975 The Asilomar Conference on Recombinant DNA is held at the Asilomar (California) Conference Center. About 140 scientists, physicians, and legal experts meet to draw up voluntary guidelines for the conduct of recombinant DNA research because of its potential for serious health and ethical issues.

1976 American venture capitalist Robert A. Swanson and Stanley N. Cohen cofound Genentech, Inc., a corporation designed to develop and market applications of rDNA technology originally discovered by Cohen and Boyer in 1973.

The NIH issues the first guidelines for rDNA research. The guidelines are based largely on recommendations from the Asilomar Conference (1975).

1977 Genentech announces the production of the first human protein manufactured by bacteria—somatostatin—a human growth hormone. The protein is the first commercial product of industrial rDNA research.

1980 The U.S. Supreme Court rules that a patent may be issued to the General Electric company for a genetically engineered bacterium invented by Indian American chemist Ananda Chakrabarty. The Court rules that the fact that the item in question is alive is not an impediment to its being patented. The decision sets a precedent for the granting of patents for all manner of engineered organisms and components of life (such as genes).

A group of Canadian researchers successfully introduces the human gene for the production of the protein interferon into a bacterium, the first time a human gene has been so transferred. The discovery makes possible the relatively simple and inexpensive production of the medically valuable protein.

American biochemist Kary Mullis invents the polymerase chain reaction (PCR), a method for making a very large number of copies of a given DNA specimen. PCR is now one of the most widely used procedures in DNA technology.

The U.S. Patent and Technology Office issues patents to Stanford University for the fundamental technology used in recombinant DNA technology. Over ensuing years, Stanford sells licenses to at least six dozen companies for the use of this technology in their own research and development programs.

1982   California researcher Stephen Lindow requests permission from the NIH to field-test an engineered bacterium that promises to make strawberries, potatoes, and other plants less vulnerable to frost damage. A year later, the NIH grants permission for the tests.

1983   Researchers locate the genetic marker for Huntington's disease on chromosome number 4, the first time a disease-causing gene has been specifically mapped in the human genome.

1984   British biochemist Sir Alec Jeffreys develops a process known as restriction fragment length polymorphism (RFLP), a technique by which DNA from different organisms can be distinguished from each other. It later becomes a powerful tool in forensic analysis.

Researchers at the University of Pennsylvania, the U.S. Department of Agriculture Agricultural Research Service, and the University of Washington create transgenic pigs, sheep, and rabbits by injecting foreign DNA into the organisms' eggs and allowing them to develop and come to term.

1985   The NIH issues the first guidelines for human gene therapy.

1986   RFLP analysis (see 1984) is used to convict Colin Pitchfork of the murder of two girls in Narborough, England. The case is the first incident in which DNA evidence is used to solve a crime.

The U.S. Department of Agriculture (USDA) authorizes the first release of a genetically engineered organism, a type of tobacco plant developed by the Agracetus company that is resistant to crown gall disease.

The U.S. Food and Drug Administration (FDA) grants approval for use of the first genetically engineered vaccine, a product made by the Chiron company for use against the hepatitis B virus.

All agencies of the federal government responsible for the oversight of DNA technology in the United States join to issue a set of guidelines, "Coordinated Framework for Regulation of Biotechnology," that clearly specifies the responsibilities assigned to the FDA, NIH, U.S. EPA, USDA, and other agencies involved in regulation.

For the first time, researchers clone a disease-producing gene, the gene that causes chronic granulomatous disease, using a method known as positional cloning. Positional cloning allows one to find a gene in the human genome without knowing anything about the protein it produces.

1987   A committee especially chartered by the U.S. Congress, the Health and Environmental Research Advisory Committee (HERAC), recommends the creation of a 15-year program to sequence the human genome (see 1989).

1988   The USDA authorizes field tests of the Calgene corporation's Flavr Savr™ genetically engineered tomato. The tomato carries a gene that suppresses the gene that causes a plant to continue ripening. The product is later approved for production and sale, but economic issues cause the company to discontinue its production in 1997.

Harvard molecular geneticists Philip Leder and Timothy Stewart are awarded the first patent for a genetically altered animal, a mouse that is highly susceptible to breast cancer, the so-called Harvard Oncomouse. The European Union later approves a patent for the same discovery in 2001, although the Canadian Supreme Court denies a similar application in 2002.

The U.S. Federal Bureau of Investigation (FBI) establishes a laboratory for the testing of DNA samples in criminal investigations.

Tommy Lee Andrews, a resident of Orlando, Florida, is convicted of rape at least partly as the result of DNA typing evidence submitted by the prosecution. The case is the first instance in the United States in which DNA evidence is accepted in a criminal case.

On the basis of DNA evidence obtained from the scene of the crime, Gary Dotson is exonerated of rape and aggravated kidnaping charges for which he was convicted in 1979.

1989 The National Center for Human Genome Research is established for the purpose of conducting and sponsoring research on the human genome and related issues. The first director of the center is James Watson.

The admissibility of DNA evidence is challenged for the first time in the United States in the case of *People v. Castro*. The court rules that DNA evidence can be used to exclude suspects, but not as evidence of guilt.

The Virginia Supreme Court confirms the death penalty for Timothy Wilson based on DNA evidence presented in his earlier trials. The case is the first instance in the United States in which DNA evidence leads to the death penalty.

1990 Researchers at the National Heart, Lung, and Blood Institute and the National Cancer Institute perform the world's first human gene therapy treatment on two children with adenosine deaminase deficiency, a severe hereditary disease. Later studies appear to show that the treatment is at least partially successful, and a similar procedure has since been used with more than two dozen patients worldwide.

The biotechnology company Calgene develops an engineered form of cotton resistant to the herbicide bromoxynil (Buctril®), used to treat broadleaf weeds in cotton (and other crop) fields.

The FBI initiates a pilot DNA database program called the Combined DNA Index System (CODIS), designed to provide an efficient system by which federal, regional, state, and local law enforcement agencies can exchange DNA typing evidence from criminal scenes.

1992 The USDA issues a policy statement indicating that food from genetically engineered plants will not be regulated any differently from conventional foods.

A team of British and American doctors develops a procedure for testing embryos for the presence of certain genetic disorders, such as hemophilia and cystic fibrosis. They remove individual cells from four- or eight-cell embryos and look for defective genes that may cause such disorders.

1993   The FDA approves the use of genetically engineered bovine somatotropin hormone (bST) for use in dairy cows to increase milk output.

Kirk Bloodsworth is exonerated by DNA typing of the 1984 murder and rape of a nine-year-old girl in Rosedale, Maryland. He is the first person on death row to be exonerated as a result of the technology.

1994   The FDA approves the marketing of the Flavr Savr™ tomato.

1995   The EPA approves the first plant genetically engineered to resist attack by pests, the New Leaf potato, developed by Monsanto.

The Ethical, Legal, and Social Issues (ELSI) division of the Human Genome Project (HGP) publishes a model Genetic Privacy Act recommended for adoption by individual states until the U.S. Congress enacts a federal law dealing with privacy issues occasioned by the use of genetic testing.

A team of researchers from Johns Hopkins University, the State University of New York in Buffalo, the National Institute of Standards and Technology, and the Institute for Genomic Research sequence the first complete genome of a freeliving organism, the bacterium *Haemophilus influenzae*. The genome consists of 1,749 genes.

1996   Initial research in the sequencing of the human genome under the Human Genome Project begins at six universities in programs sponsored by the federal government.

A joint NIH/Department of Energy Committee appointed to evaluate the ELSI program of the Human Genome Project issues a report containing recommendations about future directions of the ELSI component of the HGP.

The Office of Justice Programs at the U.S. Department of Justice publishes a report, "Convicted by Juries, Exonerated by

Science: Case Studies in the Use of DNA Evidence to Establish Innocence after Trial," that summarizes the use of DNA typing in proving the innocence of men and women previously convicted of crimes.

1997   Researchers at the Human Genome Project begin to identify specific genes responsible in full or part for a variety of human diseases, including breast cancer, Parkinson's disease, and Pendred syndrome.

HGP researchers completely sequence the first human chromosome, chromosome number 7.

The European Parliament adopts a set of rules and regulations concerning the production and sale of genetically modified foods within the European Union.

Researchers at the Roslyn Institute in Edinburgh, Scotland, clone the first mammal, a sheep given the name of Dolly, from an adult somatic cell by the process of nuclear transfer.

The Monsanto corporation introduces the first herbicide-resistant food crop, Roundup Ready® soybeans, and the first pesticide-resistant crop, Bollgard® cotton.

1999   Eighteen-year-old Jesse Gelsinger is the first person reported to have died following a gene therapy procedure. Gelsinger was being treated for ornithine transcarbamylase deficiency, a condition that is normally fatal at birth, but which he had survived because of its occurrence as the result of a mutation rather than a genetic defect.

The Forensic Science Service in the United Kingdom develops a system for analyzing very small amounts of DNA known as low copy number (LCN) DNA. The technology becomes popular among some forensic scientists, but also raises questions about the validity of the technology.

2000   President Bill Clinton issues Executive Order 13145, which prohibits discrimination in federal employment based on genetic information.

Ingo Potrykus, emeritus professor of plant sciences at the Swiss Federal Institute of Technology, and his colleagues announce the availability of golden rice, rice that has been genetically engineered to produce beta-carotene, used in the production of vitamin A. The new form of rice provides a simple and inexpensive way to prevent blindness in millions of children around the world.

2001   The FDA issues a set of guidelines for the voluntary labeling of foods produced in the United States that have been produced by some type of genetic engineering. Lacking federal legislation on this issue, these guidelines are the closest thing currently available to legal regulations on the labeling of genetically modified foods.

Two separate research groups announce the nearly complete sequencing of the human genome. One group consists of researchers in the Human Genome Project, and the second group is a privately funded group of researchers at Celera Genomics. The "nearly complete" genome consists of about 83 percent of all known human genes.

The FDA approves the use of the drug imatinib (Gleevec®) for the treatment of certain types of cancer. Imatinib is the first of a group of so- called "targeted drugs"—drugs that act on one specific gene to prevent the production of a harmful protein/ enzyme responsible for a disease. The ability to make targeted drugs is a result of scientists' new understanding of the human genome.

Researchers at Texas A&M University clone the first pet, a kitten named C.C. (for carbon copy). At a cost of about $50,000, the pet owner received an exact copy of a favorite cat that had died.

The U.S. Equal Employment Opportunity Commission (EEOC) files suit against the Burlington Northern Santa Fe Railroad to end genetic testing of some of its employees without their knowledge. The suit is settled out of court a year later. The case was the first instance in which the EEOC filed suit against genetic testing based on the Americans with Disabilities Act of 1990.

2002 The USDA creates a new agency within the Animal and Plant Health Inspection Service, the Biotechnology Regulatory Services. The mission of the service is to focus on the department's responsibilities for regulating and facilitating biotechnology.

The Human Genome Project announces the creation of the International HapMap Project, a program that allows access to information so that researchers can pursue studies on the relationship of specific genes in the human genome responsible for a number of common diseases, including asthma, cancer, diabetes, and heart disease.

Merck & Company develops a genetically engineered vaccine for use against four types of human papillomavirus (HPV), implicated in the development of cervical cancer. The vaccine is developed by recombinant DNA technology.

Researchers at the State University of New York at Stony Brook synthesize the first artificial virus, an identical copy of the virus that causes polio. They use easily obtained information and materials and call the process "very easy to do," although their work raises serious ethical concerns about the use of the technique by terrorists and the possibility of using it to create higher forms of life.

A team of British scientists discovers a gene that increases one's susceptibility to depression, raising the possibility of treating mental disorders with genetic therapy.

China's State Food and Drug Administration grants the world's first regulatory approval of a drug called Gendicine for the treatment of squamous cell head and neck cancer. The drug contains an engineered copy of the p53 gene, known for its ability to destroy cancer cells.

Researchers at the University of North Carolina develop a method for treating the genetic disorder thalassemia by repairing erroneous copies of messenger RNA rather than attempting to repair the incorrect gene responsible for the production of the mRNA.

Researchers at Emory University in Atlanta cure mice of sickle-cell anemia by transplanting corrected copies of defective genes responsible for the disease into the animals.

The National Academy of Sciences publishes a report on the environmental effects of transgenic plants, which includes a number of recommendations on the topic. The investigating committee points out that the environmental impacts of conventional farming are poorly known and should be studied, along with similar research on transgenic plants—such plants should be studied and evaluated on a case-by-case basis.

2003   Attorney General John Ashcroft announces creation of the President's DNA Initiative, a program designed to ensure that DNA typing is used to its greatest potential in solving crimes, protecting the innocent, and identifying missing persons. The initiative includes funding, training, and assistance for federal, state, and local forensic laboratories and law enforcement personnel.

In February, the FDA places a temporary hold on all human gene therapy trials that use retroviral vectors to insert genes. The ruling comes when the second of two French children treated (successfully) for X-linked severe combined immunodeficiency disease (X-SCID) develops leukemia. The FDA lifts the ban in April 2003 when it becomes clear that the technology used in the French experiments is not fundamentally flawed.

2004   A team of Japanese and American researchers develops methods for producing a transgenic animal (zebra fish), resulting from the fertilizing of conventional fish eggs by sperm that have been genetically modified in a laboratory dish. The technology holds significant progress for the production of other transgenic animals by methods that are qualitatively different from those used in the past.

The United Nations Food and Agriculture Organization issues a report, "The State of World Food and Agriculture 2004," that says that genetically engineered crops may provide a huge benefit to the millions of people around the world who do not get enough to eat every day, but that questions risks to human health and the environment from such crops that remain to be answered.

The U.S. Institute of Medicine issues a report that says that genetically modified foods pose no more health risks than foods produced by conventional methods. The report suggests that all foods should be evaluated on the basis of their properties and characteristics, and not on the basis of the method used to produce them.

The FDA approves an application by the company Genomic Health for one of its products, Oncotype DX™, for the determination of the likelihood of reoccurrence of breast cancer. The test measures the activity of certain genes known to be implicated in the development of breast cancer.

Voters in California approve Proposition 69, which requires that all convicted felons and anyone arrested for or charged with felonies and some misdemeanors be required to submit DNA samples for placement in a state DNA database. The proposition applies to both adults and juveniles. The proposition passes 62 percent to 38 percent.

2005   The United Nations General Assembly adopts a somewhat ambiguous resolution opposing cloning for the purposes of producing human life but, presumably, supporting cloning for therapeutic reasons.

The World Health Organization (WHO) issues a report, "Modern Food Biotechnology, Human Health and Development," which concludes that engineered crops can increase yield, improve food quality, and expand the diversity of foods available to consumers, leading to improved nutrition and better overall health for millions of people around the world.

2006   A public-private partnership called the Genetic Association Information Network is created to study the genetic components of six major diseases: attention deficit hyperactivity disorder, bipolar disorder, diabetic nephropathy, major depressive disorder, psoriasis, and schizophrenia.

The USDA's Center for Veterinary Biologics gives approval to Dow AgroSciences for the production of the world's first plant-made vaccine. The vaccine is used to protect chickens from Newcastle disease, a highly contagious and fatal viral infection. The production process for the vaccine makes use of plant parts only, and not complete plants, and is conducted in an enclosed facility to prevent escape of genetic components into the environment.

The USDA gives its approval to the Renessen Corporation to begin manufacture and sale of its genetically engineered corn Mavera™, which contains an added gene for the production of lysine. Mavera™ corn is more nutritious for swine and poultry than is conventional corn, improving the quality of meat and reducing the cost of production for dairy operators.

The World Trade Organization rules that the European Union's ban on genetically modified foods is illegal in light of overwhelming evidence that engineered foods pose no more of a risk to human health than do conventional foods. The ruling comes in response to a complaint filed by Argentina, Australia, Brazil, Canada, India, Mexico, New Zealand, and the United States.

Researchers at the National Cancer Institute use genetically engineered immune cells to successfully treat two patients with metastatic melanoma, an aggressive form of cancer. Fifteen other patients in the study receive no benefit from the treatment, but the study confirms the possibility of using gene therapy for the treatment of at least some forms of cancer.

2007   British scientists use gene therapy to successfully treat patients with Leber's congenital amaurosis, a genetic disorder that causes severe loss of sight and/or blindness.

2008   The U.S. Congress passes and President George W. Bush signs the Genetic Nondiscrimination Act, which prohibits discrimination in hiring, employment, and related situations on the basis of information obtained from genetic testing.

The United Nations adopts an additional protocol to its Convention on Human Rights and Biomedicine opposing discrimination based on information obtained as the result of genetic testing.

The FDA approves the use of Recothrom®, a genetically engineered product that induces blood clotting, with potentially extensive use in surgery when typical blood clotting cannot be used.

2009   The FDA announces its final ruling for the regulation of genetically engineered animals. Such animals are to be approved on a case-by-case basis by the FDA, but food products produced from them do not have to carry special labels.

2010   Researchers at the Venter Institute announce the creation of the first synthetic bacterial genome.

2011   The journal *Nature Methods* selects genome editing with engineered nucleases as its "method of the year."

2013   The state of Connecticut enacts a GMO labeling law that will take effect only when other neighboring states adopt a similar law.

The journal *Nature Methods* selects single-cell sequencing as its "method of the year."

2014   Scientists at the Scripps Research Institute and New England Biolabs report the creation of a "semi-synthetic" organism that makes use of several "unnatural base pairs," that is base pairs other than A-T and G-C normally found in natural DNA.

The Vermont legislature passes a bill requiring foods containing genetically modified components to be so labeled, the act to take effect on July 1, 2016.

2015   The United States District Court for the District of Vermont rules against the Grocery Manufacturers Association's (GMA) suit against the Vermont law requiring the labeling of genetically modified foods.

2016   Researchers at the Venter Institute announce creation of a synthetic bacterium with the smallest possible genome, consisting of 473 genes.

Brooklyn judge Mark Dwyer rules that evidence obtained from LCN DNA analysis is not admissible in a trial of two individuals accused of violent felonies because the technology is not yet widely accepted by the community of forensic scientists.

Great Britain's Human Fertilisation and Embryology Authority grants approval for researchers to conduct gene-editing experiments on human embryos.

Glossary

allele   One member of a pair or series of different forms of a gene. An allele is a specific sequence of nucleotides on a DNA molecule.

amino acid   A biochemical compound containing the amino group ($-NH_2$) and one or more carboxyl groups ($-COOH$). Proteins are made of many (dozens, hundreds, or thousands) of amino acids joined to each other.

amplification   The process by which the number of copies of DNA is increased.

analyte   A substance that is measured in a laboratory test, such as a specific hormone or enzyme.

bacteriophage   *See* phage.

base pair (bp)   Two nitrogen bases (either adenine and thymine or cytosine and guanine) joined by hydrogen bonds in a DNA molecule.

base sequence   The order in which nitrogen bases occur along a DNA molecule. Base sequence determines the protein formed by the DNA molecule.

bioballistics   A method for inserting genes into a host cell in which thin metal slivers are coated with genes and fired into the cell by some type of propulsive device.

candidate gene   A gene thought to be responsible for some specific disease.

chemical poration   A method for inserting genes into a host cell in which the cell is treated with some chemical that will produce tiny holes in the cell walls, allowing genes to be inserted into the cell more easily.

chimera   An organism produced by the combination of two distinctly different organisms, as in the insertion of human gene into a mouse or the cross-breeding of a sheep and a goat.

chromosome   A thread-like structure found in the nucleus of a cell that consists of DNA and a variety of structural proteins.

clone   An organism that is identical in its DNA composition to some other organism, produced by some form of asexual reproduction.

codon   A sequence of three adjacent nucleotides on a strand of DNA or RNA that specifies the genetic code for the synthesis of some specific amino acid.

complementary DNA (cDNA)   A single strand of DNA synthesized in the laboratory from an RNA molecule using the enzyme reverse transcriptase.

confidentiality   In the field of genetic testing, the principle that information obtained from a genetic test cannot be provided to any other person without permission of the person being tested.

deoxyribonucleic acid. *See* DNA.

deoxyribose   A pentose (five-carbon sugar) that is one of the three major components of DNA. Deoxyribose has one hydroxyl (-OH) group fewer than ribose.

DNA   A biochemical molecule consisting of two strands wrapped around each other in a helical formation in which the individual strands consist of alternating units of deoxyribose sugar and phosphate groups held together by pairs of nitrogen bases.

DNA probe   *See* probe.

DNA replication   The process by which a DNA molecule unwinds and makes an exact copy of itself.

DNA sequence   The order in which nitrogen bases occur in a DNA molecule.

DNA sequencing   Any set of procedures by which the exact order of nitrogen bases in a molecule of DNA is determined.

double helix   The geometric shape of a DNA molecule, consisting of two strands of polynucleotides wrapped around each other in a "spiral staircase" formation.

edible vaccine   A vaccine that consists of some edible food into which has been inserted one or more genes for an antigen.

electrophoresis   A laboratory procedure by which molecules are separated from each other on the basis of the rate at which they migrate in an electrical field.

electroporation   A method for inserting genes into a host cell by treating the cell with an electric shock to produce tiny holes in the cell wall, allowing genes to be inserted more easily.

enzyme   A protein that acts as a catalyst, that is, that increases the rate of a chemical reaction in the body.

eugenics   A philosophical system and field of study devoted to the belief that a population of organisms can be improved by selective breeding.

exogenous DNA   DNA that has been introduced into an organism.

exon   A segment of a DNA molecule that codes for a gene.

expressed sequence tag (EST)   A portion of a cDNA molecule that can be used to identify a gene.

gene   The unit of heredity. A gene is a segment of DNA responsible for the synthesis of messenger RNA in the production of a protein.

gene amplification   The process by which numerous copies of a segment of DNA are made. Gene amplification can occur naturally or as the result of experimental procedures.

gene mapping   The process by which the location and linear separation of genes on a chromosome are determined.

gene therapy   Any process by which scientists attempt to cure or ameliorate the effects of a genetic disease by correcting, replacing, or supplementing incorrectly functioning or non-functioning genes in an organism.

gene transfer   The process by which a gene is inserted into an organism. Gene transfer is conducted for a number of different reasons and by using a variety of technologies.

genetic code   The "instructions" carried in a DNA molecule that tells that molecule what protein to make. The genetic code consists of various combinations of three nitrogen bases.

genetic counseling   The process by which information is provided to individuals about the possible risks of genetic disorders, with discussions of alternative actions that may be available. Genetic counseling is often a critical aspect of any genetic testing or screening program.

genetic discrimination   Actions taken against an individual or a group based solely or primarily on the basis of genetic characteristics.

genetic predisposition   An individual's susceptibility to a disease because of some error in his or her genetic characteristics. Genetic predisposition does not mean one necessarily develops that disease; only that he or she has the genetic characteristics for it.

genetic screening   Genetic testing on a group of people to determine their susceptibility to a particular disease.

genetic testing   The analysis of an individual's genetic characteristics to determine the likelihood that he or she may develop a disease or to confirm a diagnosis of such a disease.

genetically engineered food   *See* genetically modified food.

genetically modified food   A food or food ingredient consisting of or containing one or more genetically modified organisms or the product(s) of such organisms.

genome   The completed genetic composition of an organism, often expressed in terms of the nitrogen base pairs that occur in the organism's DNA.

genomics   The study of genes and their function.

human gene therapy   *See* gene therapy.

in vitro   A procedure carried out in the laboratory as opposed to within a living organism (from the Latin expression for "within glass").

in vivo   A procedure that takes place within a living organism, as opposed to in a laboratory setting (from the Latin expression for "within life").

informed consent   The principle that one agrees to take part in some activity with the proviso that he or she completely understands the risks and benefits of that activity.

insertion   In genetics, a type of mutation that occurs when some foreign material (one or more base pairs) is inserted into the DNA sequence of a gene, often disrupting that gene and the function it normally has.

intron   A DNA sequence that separates the coding parts of a gene. Introns are transcribed into messenger RNA, but are cut out before synthesis of a protein from the mRNA.

junk DNA   DNA that has no known coding function.

knockout   The process of deactivating a gene, usually for the purpose of studying the properties and function of that gene.

laser poration   A method for inserting genes into a host cell by treating the cell with a laser beam, which produces tiny holes in the cell wall, making it easier to insert genes into the cell.

ligase   An enzyme that catalyzes the formation of hydrogen bonds between two DNA or RNA fragments.

messenger RNA   *See* mRNA.

microinjection   A process for introducing DNA into a host cell using a fine-tipped pipet.

mitochondrial DNA (mDNA)   DNA found in the mitochondria in the cytoplasm of a cell.

mRNA   A form of RNA that transmits information stored in a DNA molecule to ribosomes, where it is used in the synthesis of proteins.

nitrogen base   A biochemical compound related to one of two basic compounds, purine and pyrimidine, found in DNA and RNA. Adenine, cytosine, guanine, and thymine are the nitrogen bases in DNA, while adenine, cytosine, guanine, and uracil occur in RNA.

nuclear transfer   The process by which the nucleus of a cell is removed from one cell and transferred to a second cell, whose own nucleus has already been removed.

nucleic acids   A family of organic compounds that consist of three basic parts: a phosphate group, a sugar (ribose or deoxyribose), and a small number of nitrogen bases.

nucleoside   A subunit of a nucleic acid consisting of one nitrogen base and one sugar (ribose or deoxyribose) attached to each other.

nucleotide   The basic unit of a DNA molecule, consisting of a phosphate group, the sugar deoxyribose, and one of four nitrogen bases: adenine, cytosine, guanine, or thymine.

penetrance   The tendency of a particular genotype to be expressed as a particular phenotype. If the genotype always produces the characteristic phenotype, the genotype is said to be completely penetrant. If it is expressed only some of the time, the genotype is said to be incompletely penetrant.

phage   Shorthand for *bacteriophage*, a virus that attacks bacterial cells, takes control of the cell's system of reproduction, and makes copies of itself.

pharmacogenics   The study of the way in which a person's genetic makeup affects his or her reaction to drugs.

pharming   The use of DNA technology to engineer a plant or animal in order to make a commercially useful product.

plasmid   An extra-chromosomal double-stranded circular piece of DNA found in bacteria and protozoa.

polymorphism (genetic)   A frequently occurring variation in the nucleotide sequence in a DNA molecule. Genetic polymorphisms are responsible for the production of polymorphic proteins.

positive predictor value (PPV)   The likelihood that a person will develop a disease for which he or she is being tested.

precautionary principle   The concept that regulatory agencies may be justified in taking regulatory actions even when some scientific uncertainty exists as to the possible risks and consequences of some given practice.

probe   A single-stranded DNA or RNA molecule prepared synthetically, with a specific base sequence, used to detect a DNA or RNA segment with the complementary base pattern. The probe also carries some time of label, such as a radioactive isotope, that allows its location to be tracked.

proteomics   The study of the full set of proteins produced by some given genome.

recombinant DNA (rDNA)   A synthetic DNA molecule produced by the insertion of some foreign DNA segment into a host molecule.

restriction endonuclease   *See* restriction enzyme.

restriction enzyme   An enzyme that recognizes specific base segments in a DNA molecule and then cuts those segments at specific positions.

ribonucleic acid   *See* RNA.

RNA   A biochemical compound that contains the sugar ribose, phosphate groups, and four nitrogen bases: adenosine, cytosine, guanine, and uracil. Various forms of RNA exist in cells, each form with a specific function.

somatic cell gene therapy   The insertion of genetic material into a cell for some therapeutic purpose. Since the cell is somatic, the gene modification cannot be passed to offspring.

Southern blot   A common laboratory procedure used to determine the base sequence in a DNA fragment.

traceability tag   A piece of DNA added to a genetically modified food product that has no effect on human health, the environment, or the organism into which it is inserted, but that provides a recognizable "address" of the company that made the product.

transcription   The synthesis of a messenger RNA molecule from the genetic code (base sequence) stored in a DNA molecule.

transfection   The introduction of DNA into a host cell.

transgenic organism   An organism formed by the insertion of some foreign DNA into its genome.

translation   The process by which the genetic information (base sequence) stored in a messenger RNA molecule is used to synthesize a protein on a ribosome.

vector   Any material, such as a virus or plasmid, used to introduce DNA into a host.

Index

About the Author

David E. Newton holds an associate's degree in science from Grand Rapids (Michigan) Junior College, a BA in chemistry (with high distinction), an MA in education from the University of Michigan, and an EdD in science education from Harvard University. He is the author of more than 400 textbooks, encyclopedias, resource books, research manuals, laboratory manuals, trade books, and other educational materials. He taught mathematics, chemistry, and physical science in Grand Rapids, Michigan, for 13 years; was professor of chemistry and physics at Salem State College in Massachusetts for 15 years; and was adjunct professor in the College of Professional Studies at the University of San Francisco for ten years.

The author's previous books for ABC-CLIO include *Global Warming* (1993), *Gay and Lesbian Rights-A Resource Handbook* (1994, 2009), *The Ozone Dilemma* (1995), *Violence and the Mass Media* (1996), *Environmental Justice* (1996, 2009), *Encyclopedia of Cryptology* (1997), *Social Issues in Science and Technology: An Encyclopedia* (1999), *DNA Technology* (2009), *Sexual Health* (2010), *The Animal Experimentation Debate* (2013), *Marijuana* (2013), *World Energy Crisis* (2013), *Steroids and Doping in Sports* (2014), *GMO Food* (2014), *Science and Political Controversy* (2014), *Wind Energy* (2015), *Fracking* (2015), *Solar Energy* (2015), *Youth Substance Abuse* (2016), and *Global Water Crisis* (2016). His other recent books include *Physics: Oryx Frontiers of Science Series* (2000), *Sick!* (4 volumes) (2000), *Science, Technology, and Society: The Impact of Science in the 19th Century* (2 volumes; 2001), *Encyclopedia of Fire* (2002), *Molecular Nanotechnology: Oryx Frontiers of Science Series* (2002), *Encyclopedia of Water* (2003), *Encyclopedia of Air* (2004), *The New Chemistry* (6 volumes; 2007), *Nuclear Power* (2005), *Stem Cell Research* (2006), *Latinos in the Sciences, Math, and Professions* (2007), and *DNA Evidence and Forensic Science* (2008). He has also been an updating and consulting editor on a number of books and reference works, including *Chemical Compounds* (2005), *Chemical Elements* (2006), *Encyclopedia of Endangered Species* (2006), *World of Mathematics* (2006), *World of Chemistry* (2006), *World of Health* (2006), *UXL Encyclopedia of Science* (2007), *Alternative Medicine* (2008), *Grzimek's Animal Life Encyclopedia* (2009), *Community Health* (2009), *Genetic Medicine* (2009), *The Gale Encyclopedia of Medicine* (2010–2011), *The Gale Encyclopedia of Alternative Medicine* (2013), *Discoveries in Modern Science: Exploration, Invention, and Technology* (2013–2014), and *Science in Context* (2013–2014).